NUMBERS, MEASURES, AND THE TRANSFER OF GOODS IN PREHISTORY

WARSAW STUDIES IN ARCHAEOLOGY

VOLUME 6

Published volumes in this series are listed at the back of the book.

Numbers, Measures, and the Transfer of Goods in Prehistory

Edited by

PHILIPPE DELLA CASA and ALEKSANDER DZBYŃSKI

BREPOLS

British Library Cataloguing in Publication Data
A catalogue record for this book is available from the British Library

D/2024/0095/56
ISBN 978-2-503-61068-9
e-ISBN 978-2-503-61069-6
ISSN 2953-1802
e-ISSN 2953-1810

DOI 10.1484/M.WSA-EB.5.136746

Printed in the EU on acid-free paper

Table of Contents

List of Illustrations

2. Between East and West – Aleksander Dzbyński

3. Feasting and Burial Rites in Pre-Hispanic Honduras – Franziska Fecher

4. Development and Functioning of Numeral Systems – Jadranka Gvozdanović

5. Some Further Thoughts on Commodity and Gift Exchange in Tribal Societies – Jürg Helbling

6. The Early Bronze Age Rib Ingot Hoard from Oberding (Upper Bavaria) – Sabrina Kutscher

7. Numbers and Measures in the Iberian Peninsula during the Iron Age – Thibaud Poigt and Coline Ruiz Darasse

8. Highly Composite Numbers and the Early Use of Weights – Lorenz Rahmstorf

Foreword

This volume contains papers from two workshops held in 2020 and 2021 at the University of Zurich. The first workshop entitled 'Number and Measure in Prehistory' took place on 8–9 May 2020. The second workshop was held on 26 May 2021 and was entitled 'Gift, Exchange, and Goods Transfer in Prehistory'. The shared element of the two topics is the ways in which value was measured in pre-industrial societies, oscillating between the phenomena of gift, debt, and money as they were shaped long ago.

The pandemic, however, thwarted our plans. Instead of meeting live, as we had originally planned, we had to move the meetings to a well-known digital platform. Despite the limitations of these virtual meetings, we believe that they were very fruitful and brought significant benefits to all participants. It should be noted at this point that there were three parties actively involved in the workshop: the invited guests, Philippe Della Casa and Aleksander Dzbyński as organizers, and students of the Institute of Archaeology at the University of Zurich. For the students, the workshop was part of two curricular cycles of courses carried out at the Institute of Archaeology of the UZH as part of a didactic and scientific project initiated by Aleksander Dzbyński; Philippe Della Casa as head of the Institute gave full organizational and substantive support to the project, extending it over two years of fruitful cooperation.

The issue of numerosity and measurement in prehistory, which constitutes the first part of this volume, was until twenty years ago considered as speculative, unconvincing, and marginal. With time, however, the amount of sources and the number of researchers who directly or indirectly contributed to the development of this very interdisciplinary issue began to grow. We are dealing here with links between archaeology, ethnology, anthropology, cognitive sciences, history of mathematics, philosophy, and philology. Each of these disciplines has developed its own research apparatus, its own terminology, methodology, and its own research questions. Nevertheless, at a basic level they are all concerned with researching and asking themselves questions about the same process: How did humanity develop that distinctive way of thinking we call mathematics, through which it was able to develop technology and civilization? We all know that the central role in this process falls to material culture, hence the important role of archaeologists, as already pointed out by one of our esteemed speakers in a 2017 article asking: 'How can archaeological evidence improve our understanding of the origins of numbers?' (Núñez 2017, 421). It may seem surprising, to say the least (but logical upon further reflection), that the essential part of the common research programme outlined above was emphatically formulated by philosophy almost a century ago. In his last unfinished book from 1936: *Die Krisis der europäischen Wissenschaften und die transzendentale Phänomenologie: Eine Einleitung in die phänomenologische Philosophie* the German philosopher Edmund Husserl wrote:

> The geometry of idealities [ideal shapes, author's note] was preceded by the practical art of field measurement, which knew nothing of idealities. But such pre-metric achievement was the foundation of meaning for geometry, the foundation for the great invention of idealization: included in this was the invention of the ideal world of geometry or the methodology of objectifying the determination of idealities through the constructions creating 'mathematical existence'. (Husserl 1954, VI, 42)

In this passage, too, we have a reference to the world of pre-metric and pre-mathematical values, the world studied by the disciplines listed above. It must be remembered here that Husserl was both a mathematician and a philosopher.

The issue of gift, exchange, and goods transfer in prehistory, which constitutes the second part of this volume, was proposed by Philippe Della Casa. The topic emerged from a renewed interest in the question of how materialist transfer and exchange in and between prehistoric societies formed and evolved, and as a wish to overcome overly simplistic approaches to economic transactions as reflected by the indiscriminate use of the term 'trade' in archaeological literature. For trade — if at all a suitable word to describe the exchange of goods in pre-mercantilist societies — is evidently only one, and obviously not the primary instrument to establish and maintain contacts and networks among related communities, as a comparative anthropological approach to the theme is able to demonstrate (cf. the contributions by Darmangeat and Helbling in this volume).

We invited a group of distinguished researchers to our workshops, representing various fields of science and humanities, thus creating, we believe, a truly interdisciplinary discussion forum. Some of them decided to submit their papers for publication, others preferred only to give oral presentations. We would like to thank all of them equally for their participation, contribution to the common discussion, and exchange of ideas.

We would like now to introduce the contributions to the volume and show how they interact with each other.

In the initial article, Christophe Darmangeat expands upon Alain Testart's model and asserts its superior alignment with ethnographic evidence concerning regulations governing the exchange of goods in contemporary non-state societies. Along with Testart, he proposes to delimit these concepts by using the criterion of legal exigibility. This analysis, which also reveals the existence of other forms of simple transfers, can be deepened by exploring complex (combined or indeterminate) forms. Darmangeat further contends that this model provides archaeologists with a wider range of interpretive possibilities regarding the underlying social processes behind the distribution of these artefacts, as opposed to the original model proposed by Marcel Mauss. Within the book's context, the article harmonizes effectively with Helbling's and indirectly Seaford's contributions, while aiming to establish a precise terminology for transfers within non-state societies, avoiding the pitfall of oversimplified classifications such as gift versus exchange or gift versus commodity.

The article by Aleksander Dzbyński is a summarizing attempt to interpret the main development paths of Europe and the Near East. On the one side, we have the Near Eastern phase of number and measure development constituted by artefacts such as tokens, that led in the subsequent phase to the establishment of full numeracy, literacy, and centralized power. In Europe, however, we see a strong current of early metal production that introduces numbers and measures, as it seems, by another means of cognitive conceptualization. The article sets out with theoretical considerations that are at this point not yet grounded in the discussion of empirical examples. It is foremost the phenomenology of Edmund Husserl that is explored in this text by introducing his general idea that the development of mathematics must have been moved from the real praxis to the ideal praxis, where the real praxis means basically various anthropomorphic ways of measuring and the ideal praxis means mathematics. The article demonstrates, and explains theoretically, how difficult the seemingly trivial concept of 'weight' and the practice of measuring weight really is and that historically it must have been introduced after tangible, apparent, and anthropomorphic linear measures were in place. The article resonates well with the findings of Sabrina Kutscher, which are presented in this volume showing that the Early Bronze Age ingots of Oberding were not standard weights but also shapes as she puts it. This is a conclusion that we have to take under consideration for our future studies.

The contribution of Franziska Fecher about feasting in pre-Hispanic Honduras tells a story about archaeological remains of gift and exchange. The article focuses on archaeological evidence of feasts, and puts them in continuity with early colonial records and contemporary ethnographic evidence. According to her findings, an integral part of these feasts was the consumption of food and drink, and the food remains were deposited in the vicinity of the burials. These festivities were carried out repeatedly, resulting in the accumulation of a dense concentration of ceramics, bones, and other objects over time. The article shows some material evidence in the discussion about gift and exchange, and as such it resonates well with the more theoretical contributions of Darmangeat and Helbling in this volume.

The article of Jadranka Gvozdanovic focuses on historical and prehistorical changes in numerical systems in linguistic contact situations and proposes explanations for the partial replacement of numerical systems in languages of hunter-gatherers who came into contact with agrarian societies. The author shows inter alia that the spread of numerical systems associated with the Indo-European language group confirms that of population movements as reconstructed archaeologically and genetically. The paper also shows how the analysis of number concepts and numeral forms can be indicative of prehistoric contacts and migrations. Her article fits very well with other articles in the volume reconstructing the cultural development of calculating and measuring systems. In particular, it resonates well with the text of Aleksander Dzbyński, which suggests that in the Eneolithic a new metrological vocabulary concerning the use, exchange, and interpersonal communication of the so-called macrolithic tools may have arisen, binding together material culture, language, and mathematics.

Jürg Helbling presents a very interesting article that demonstrates on a broad basis of ethnographic literature that in non-state or tribal societies where gift exchange is used to regulate alliances among, and status relations within groups, there exists also a more utilitarian mode of exchange based on division of labour that is often regulated by customary rates of exchange. He argues that commodity exchange, which in simple societies often takes the form of barter trade, is an economic activity in which goods are exchanged at a certain price and for mutual benefit of the exchange partners. He uses concepts of formal economics to describe these exchanges as 'commodity exchange' and the rates of exchange as 'prices' that reflect a balance of quantitative 'values' ascribed to the exchanged 'commodities'. With this vocabulary and the universalist assumptions behind them the article takes a theoretical position opposed to that of Richard Seaford in this volume. On the other hand, it enters into conversation with Christophe Darmangeat's contribution as it refers in part to the same literature. Together the three articles testify to the wide variety of competing concepts and theories in economic anthropology. In its substance the article is an important contribution for this volume as it furthers the understanding of the diversity of forms of exchange human societies without states have developed to both express relationships and satisfy material needs — and often in one and the same transaction.

The next article by Sabrina Kutcher is a very precise description and analysis of the fascinating rib ingot hoard from Oberding which is one the best documented Early Bronze Age hoards. The careful excavation *en bloc* and non-invasive computer tomography of the hoard allowed the author to show that the ingots were mostly tied together in bundles of ten and grouped in larger units of ten bundles, and that slightly lighter and slightly heavier ingots were combined to achieve equal total bundle group weights of 10.5 kg. Extensive preliminary work, especially an elaborate restoration and first metallurgical analyses, have already produced important insights that underline the outstanding importance of this find. Especially interesting, in the context of this volume and the preceding workshop, are the conclusions drawn from Kutcher's article. She points to the fact that the ingots were not precisely weighed. Nevertheless, they were aiming to produce ingots of similar length (30 cm ± 2–4 cm), and she proposes that there was a standardization of shape in play. This corresponds very well to the findings presented by Aleksander Dzbyński in this volume, who suggested that there was first of all the recognition of length and shape in terms of metrological perception until the Bronze Age, before the introduction of weighing systems. These similar but independent voices are a rare but fascinating feature in archaeology that indicate further discussions on the matter.

The article by Thibaud Poigt and Coline Ruiz Darasse presents an insightful discussion about Iberian weight measure systems and evidence of number notation systems in the Iron Age. The article shows that both the balance weight finds of the archaeological record and the as yet not fully deciphered Iberian script point to the widespread use of a combined binary, quinary, and decimal counting system. The article aims to provide a brief and non-exhaustive overview of what we know about numbers and measures in one of the best examples of late prehistory in western Europe: the Iberian Peninsula. With the use of a homogeneous type of weights from the Late Bronze Age to the Roman period (a period of about one millennium) and the use of written numerical and metrological sequences, this region constitutes a very good indicator of the practices related to numbers and measures during a large period of time. The authors conclude that the practice of weighing with regionally standardized weights constructed around a basic unit and multiples and fractions of it was in use long before numbers were written. In our opinion, this contribution to the volume resonates well with those of Rahmstorf, Kutscher, and Dzbyński and can be read as such, but not only.

The next article, by Lorenz Rahmstorf about highly composite numbers and the early use of weights makes the compelling case that the Mesopotamian hexagesimal counting system became dominant during the third millennium BP when weight systems were based on it. The optimal divisibility of sixty proved highly productive when weights had to be combined to measure exact amounts efficiently. The analysis of weight finds proves that the weight system indeed made use of the high divisibility of sixty. The author states that the Sumerians used the excellent advantages of the highly composite number sixty for metrology in order to achieve maximal divisibility when weighing out precise amounts of products like metals. The article is well and convincingly written and fits very well into the volume.

The last contribution comes from Richard Seaford who is the author of a fascinating book: *Money and the Early Greek Mind*. His approach to the idea of money and mathematics is truly diverse, it is both literary, anthropological, and historical. It is a very interesting and in the context of this volume very pertinent article that reflects on the ancient function of numbers in ritual and on the role money played in the rise of a quantitative concept of value and in the overall mathematization of cosmology in ancient Greece, as well as on the merging of these two functions. The paper examines numbers in Greek society from the eighth to fifth centuries BC, mainly as reflected in Homer, pre-Socratic cosmology, Aeschylus's *Oresteia*, and Plato. It focuses on two functions of numbers, both of which are conducive to social integration: on the one hand the creation of traditional memorable forms, which can have a co-ordinating function (notably in ritual), and on the other hand the expression of equivalence, notably in commerce. In our opinion, the article is the perfect way to close the volume merely by the very fact that ancient Greek society accumulated all of the phenomena discussed here and during our workshops.

With great sadness we learned of the death of Richard Seaford during the last stage of preparation of the book. His contribution takes on this special character that is reserved for those who by the act of fate have been prematurely excluded from scientific discourse. We regret very much that we could not finish the book with all participants on board. At the same time, we feel very honoured that we can give Richard the space for his utterly original, deeply thought-provoking, and inspiring contribution.

Now we would like to invite you to read!

Aleksander Dzbyński and Philippe Della Casa

Works Cited

Husserl, Edmund. 1954. *Die Krisis der europäischen Wissenschaften und die transzendentale Phänomenologie: Eine Einleitung in die phänomenologische Philosophie*, Husserliana, 6 (The Hague: Martinus Nijhoff) <http://philotextes.info/spip/IMG/pdf/husserl_krisis.pdf> [accessed 1 February 2024].

Núñez, Rafael E. 2017. 'Is There Really an Evolved Capacity for Number?', *Trends in Cognitive Sciences*, 21.6: 409–24 <https://doi.org/10.1016/j.tics.2017.03.005>.

Seaford, Richard. 2004. *Money and the Early Greek Mind: Homer, Philosophy, Tragedy* (Cambridge: Cambridge University Press).

CHRISTOPHE DARMANGEAT

1. How to Open the Gift Box?

Archaeology and the Transfer of Goods

ABSTRACT When it seeks to understand the way in which the goods it uncovers were transferred, archaeology is entirely dependent on a theory and a typology developed by social anthropology. In contrast to the traditional approach dating back to Marcel Mauss, which considers that the whole of the 'primitive' world is marked by an undifferentiation between gift and exchange, Alain Testart has proposed, on the contrary, to rigorously delimit these concepts by using the criterion of legal exigibility. This analysis, which also reveals the existence of other forms of simple transfers, can be deepened by exploring complex (combined or indeterminate) forms.

In the end, archaeology, because it approaches social relations from their material traces, cannot hope, in the absence of writing, to reconstruct directly the way in which goods were owned and transferred. This knowledge can nevertheless be approached insofar as archaeology can provide the elements that make it possible to identify a social type and, in turn, the predominant modes of transfer that prevailed within this type for a given category of goods.

KEYWORDS Gift, exchange, Alain Testart, Marcel Mauss, transfer of goods

Introduction

That rights over goods and their transfers constitute one of the key dimensions around which human societies are organized is obvious. Yet, strange as it may seem, on neither of these two aspects has the science of these societies forged a consensual basis of analysis. No general comparative study of property rights exists, and it can even be said that no serious attempt has ever been made in this direction. As far as transfers are concerned, the situation is hardly less critical. In fact, for decades, writings about societies distant from our own, either by their structure or by their epoch, have most often been marked by a conceptual blurring on this point. Since the founding work of Marcel Mauss ([1924] 2004), the concepts of gift and exchange, in particular, never being rigorously defined, seem sometimes to oppose each other, sometimes to intermingle to the point of becoming indistinguishable. Thus, when attempting to interpret the circulation of objects in distant times, archaeologists face a double difficulty. The first, inherent to their discipline, is that objects do not preserve any trace of the rights that humans have exercised over them, nor of the manner in which these rights may have changed holders. This problem, by itself, is already considerable. But in the absence of a clear definition of the various ways in which rights in property are transferred, it has no chance of finding a satisfactory solution. To use a metaphor, recognizing animals in the night is a challenge; but to succeed in doing so without knowing exactly what the different

Christophe Darmangeat Université Paris Cité, LADYSS. Christophe Darmangeat is a social anthropologist and lecturer at the Université Paris Cité. He has published several articles and books in French and English about the origin of male domination, law, and warfare in hunter-gatherer societies, or the emergence of wealth. His blog (in French): lahuttedesclasses.net.

Number and Measure in Prehistory: Gift, Exchange, and the Transfer of Goods, ed. by Philippe Della Casa and Aleksander Dzbyński, WSA, 6 (Turnhout, 2024), pp. 15–23.
BREPOLS PUBLISHERS
DOI 10.1484/M.WSA-EB.5.137126

species look like and what their characteristics are is utterly impossible.

The following article is an extension of the work of Alain Testart, who developed a criterion for classifying transfers of goods and thus identified the different fundamental types (Testart 2001; 2007). We will therefore begin by recalling the main contributions of this analysis. However, this is only the first step in a long process: social reality is infinitely more complex than simple 'pure' types would suggest. It is therefore essential to understand the way in which the latter combine to produce the former. Finally, perhaps the most difficult task remains: to determine the extent to which the transfer of goods says something about social relations, and the contribution of this analysis to archaeology.

Types of Transfers of Goods: A Classification

Gift versus Exchange

The starting point of Testart's critique is to rigorously define gift and exchange — and thus to differentiate them from each other. We know that in Mauss's text, gift and exchange are so confused that the author uses a number of expressions such as 'contractual gifts' (Mauss [1924] 2004, 10–11), the 'exchange through gifts' (Mauss [1924] 2004, 16, 23), the 'exchange-gift' (Mauss [1924] 2004, 29, 59), the 'gift-through-exchange' (Mauss [1924] 2004, 37), the 'remuneration-cum-gift' (Mauss [1924] 2004, 39), the 'obligation to give' (Mauss [1924] 2004, 16, 17, 50), the 'obligatory and voluntary exchanges of gifts' (Mauss [1924] 2004, 19), or again gifts destined 'to paying' (Mauss [1924] 2004, 94). In substance — and this is even the subtitle of the work — the gift is seen as 'the form and reason for exchange in archaic societies'. In the eyes of Alain Testart, this is a double error, both ethnographic and conceptual. Ethnographic, because the will to encompass the whole of non-state societies in a unique approach, as if within them, the transfers of goods uniformly obeyed the same rules, does not correspond to the diversity of the social types observed, and that science must precisely identify. But Mauss's approach also reflects a conceptual error, by forging oxymorons, 'logical monstrosities of the same nature as a circle-square' (Testart 2001, 742), which intermingle concepts that should, on the contrary, be carefully delimited.

The first step is to identify the pitfalls of the current vocabulary, whose imprecision is a factor of confusion. To begin with, the act of 'giving' does not necessarily mean that one is making a 'gift'. When we are at the store and we ask the shopkeeper to 'give' us a loaf of bread, we know that this transfer is not free, but that it implies a payment — and that the absence of this has many consequences. In other words, if the baker 'gives' us a loaf of bread, he does not make a 'gift' to us. The situation with respect to exchange is equally unclear, since the term has two very different meanings. In its broadest sense, exchange is purely descriptive, and expresses the idea that two transfers are made in opposite directions between two entities: this is the case, for example, in physics, of gas exchanges. It is in this sense, but only in this sense, that we can speak of an exchange of gifts, as when at the Christmas party, gifts are passed between participants and each is both giver and receiver. But exchange also has a much narrower meaning — and it is the latter that must be used in sociology: it then designates a situation where the two transfers are not only reciprocal, but where they are also the necessary condition of each other. This is exactly what happens in the bakery: the baker transfers the baguette to the customer if, and only if, the customer transfers money to him — and vice versa.

What differentiates the gift from the exchange is not, therefore, the presence or absence of a transfer in the opposite direction: if the exchange cannot, by definition, be one-way, on the other hand, a gift can very well be accompanied by a counter-gift. It is even less the simultaneous or deferred character of the reciprocal transfer: just like the counterpart of an exchange, the counter-gift can intervene at the same time as the gift (in the case of Christmas presents) or be deferred (when one returns an invitation to dinner), without this changing anything about its nature. The difference between gift and exchange lies entirely in the exigible character — in the legal sense of the term — of the counter-transference. In making a gift, one can expect and hope for a counter-gift — the counter-gift may even represent the sole motivation for making the gift. But, crucially, if there is a breach, there is no legal way to enforce this counter-gift. If the owner of a restaurant where one has dined without paying can go to court to enforce his or her right, there is no way to do the same with a friend who has not returned a restaurant invitation. Once again, therefore, the ambiguity of the term often used by Mauss, that of obligation (to give, to receive or to return), must be noted. Indeed, it is imperative to differentiate between a purely moral obligation, which can at most only be translated into social pressure, and a legal obligation, which opens the possibility for the injured party to appeal to legitimate force in order to assert his rights. For the transfer of goods, to speak of obligation in general, without making this distinction, is thus to deny oneself the possibility of thinking about the profound difference between gift and exchange. Would one object that in a society without a state, the distinction loses its relevance, the notions of right and legitimate force becoming pointless? This would be a profound error. If, in stateless societies, law is not written, and its implementation is not necessarily the work of formal and specific organs, no society exists that ignores any form of law and, consequently, of a definition of the framework of the legitimate exercise of force. This in no way closes the possibility that there may be transfers that are at the limit of exigibility, and which would therefore raise objective difficulties of classification — we shall return to this point.

Other Types of Transfers

The proposed definition of gift and exchange does not exhaust the possible range of transfers; first, because there is a third type of transfer, which is neither exchange nor gift. Secondly, because within gift and exchange, one can usefully draw additional distinctions. Let us begin with these: the gift is classically associated with the counter-gift, that is, the gift that comes in response to a previous gift, made in the opposite direction. Again, this double movement constitutes an exchange only in an extremely broad sense, insufficiently precise for sociological analysis. Strictly speaking, each of these two transfers does not constitute the condition that makes the other exigible; in other words, giving a counter-gift may be a moral obligation but, by definition, not a legal one. As regards exchange, it is necessary to distinguish free exchange, the most common form, from forms in which one of the two co-exchangers is forced to make the transfer by the other (or possibly by a third party): think, for example, of our modern expropriations and nationalizations, where the state compensates the owner from whom it buys his property.

There remain, therefore, transfers that are neither gifts nor exchanges. These are those which derive from a legal obligation — they are therefore exigible. But, unlike exchange, this exigibility does not arise from a transfer in the opposite direction. The first example that comes to mind is undoubtedly that of modern taxes, which are inseparable from the state; but one could just as easily cite payments of a judicial nature (damages or fines), which also exist in many non-state societies. These 'transfers of the third type', or 't3t', as Alain Testart called them, form the last major category of legitimate transfers, which has largely passed under the radar of a social anthropology focused on gift and exchange.

A final problem concerns the appropriateness of including all illegitimate transfers — including looting, or the trade in illicit products — in the classification. Following the paths opened up by F. Athané (2011), and in particular his concept of 'community of reference', I attempted to show how illegitimate transfers could be the subject of a classification that closely mirrored that of legitimate transfers (Darmangeat 2016). Such an analysis seems a priori to contravene the principles laid down by A. Testart — the criterion of exigibility applies, at first sight, only to legitimate transfers. On the contrary, it proves to be not only consistent with his own, but also very useful for understanding situations where, in the absence of a state, the legitimacy of a transfer can be assessed differently within the same society. However, we shall leave aside this difficult problem here, and stick to the most fundamental aspects of the analysis.

The above can be summarized with the following diagram:[1]

Typology of legitimate transfers

I. Non-exigible transfer
   - because of a non-exigible reverse transfer: ***counter-gift***
   - for another reason: ***gift***

II. Exigible transfer
   - because of a reverse transfer:
     - both parties entered freely into the relationship: ***free exchange***
     - either party entered into the exchange relationship as a result of a legal obligation: ***mandatory exchange***
   - for another reason: ***t3t***

Complex Forms

The above analysis identifies, so to speak, the elementary forms of transfer. It thus constitutes an indispensable, but not sufficient, step in the analysis: reality is not limited to these forms, and also includes more complex phenomena. These phenomena proceed from the elementary forms in two ways: combination and indeterminacy.

Combination

A transfer can be called combined when it mixes, in an inseparable way, several elementary forms. A common case, in our own world, is that of a sale at a friendly price: the seller's choice not to demand an equivalent value for the object transferred corresponds to his will to make a gift while concealing it, for one reason or another, in the form of an exchange.

1 This presentation appears to us to represent a step forward compared to the one proposed in Darmangeat (2016, 26): it rectifies some problematic formulations and moreover explicitly brings out the case of the counter-gift.

Such combinations of forms are commonly found in all societies: for example, in the institution known as *tee*, practised by the Enga-speaking people of New Guinea, which is related to what is known as 'ceremonial exchanges'. According to the detailed description provided by Daryl K. Feil (1980; 1984), the *tee* includes three specifically named transfers. The cycle is inaugurated by the *saandi*. This is most often done at the request of the recipient, who asks for certain goods — primarily pigs — from one of his exchange partners. If the individual requested cannot or does not wish to respond to the request himself, he may turn to one of his own partners (and so on). Sometime later, in a public and formalized manner, the recipients of *saandi* engage in a reverse transfer, called a *tee-pingi*, normally of equal or greater value. Finally, for each live pig received and slaughtered, each recipient of a *tee-pingi* returns one or more shares of cooked pork, called *yae*.

What is the nature of the *tee-pingi*, this counter-transfer that comes as a response to the *saandi*? In order to know whether it constitutes a (counter)gift or one of the two terms of an exchange, it is necessary to determine if it is exigible. The analysis suggests that the answer to this question is positive, but only for a portion of the transfer, that which represents an amount equivalent to or slightly greater than the value of the *saandi*. Indeed, not only does 'failure to return pork threatens the [exchange] relationship and unless it is a strong one, it cannot be smoothed over' (Feil 1984, 50, see also p. 213), but such a failure gives rise to complaints, to 'bitter quarrels, perhaps even fights' (Feil 1984, 206). The information according to which the modern courts declared themselves incompetent to judge these matters (Feil 1984, 207) is decisive: it indicates that the judicial bodies were seized of them, which meant that in the eyes of the protagonists, these disputes fell indeed within the legal domain.

The particularly high increment to which some *tee-pingis* gave rise, on the other hand, lies undoubtedly in the realm of non-exigency and therefore of gift. But 'while a profitable increase is always hoped for, a return based on Tombema equivalences is acceptable and sufficient to sustain the exchange relationship' (Feil 1984, 39). The additional *tee-pingi*, precisely because it is optional and not required, is the mark of generosity — which a local proverb expresses in the following way: 'A man with a good heart adds a pig and gives' (Feil 1984, 57). Who would call someone generous who is merely fulfilling an obligation?

In the Tombema Enga *tee*, exchange and gift are intertwined in a way that is both intimate and original. Firstly, because there is no apparent difference between exchange and gift, and the transition from one to the other is seamless. Then, because the most successful exchange is the one that gives rise, in the same movement, to a gift. Thus the gift is not only organically associated with the exchange: it constitutes its supreme goal.

Also in the highlands of New Guinea, among the Mendi this time, another type of combination is found, that of t3t and mandatory exchange. As in many other societies in this region, certain social obligations (linked in particular to marriages or deaths) required compulsory payments in goods. The originality of Mendi society was that some of these payments took the form of a double transfer: the party who was ultimately to receive the payment began by handing over a set of goods to the other. The latter was then obliged, in a second phase, to repay with increments:

> Kowar payments are obligatorily made in honor of deceased clansmen and in-married women to their matrilateral kin. Maternal relatives of the deceased make a private intiatory payment [...]. A public event, the *kowar* itself formally repays the *topowe* initiatory payment, but must also include a substantial increment, called *nopae*, in honor of the dead person. Similarly, in the context of large-scale death compensation payments (*ol tenga*) during which scores of valuables change hands between members of allied clans, *nopae* refers to pigs (or other wealth) given above the repayment of initiatory gifts in honor of men and women who died as a result of intergroup hostilities. (Lederman 1986, 149)

Despite the inappropriate use of the term 'gift' in the last sentence, the description is crystal clear, and shows how this custom constituted both an obligatory exchange in its form and a t3t in its content.

Indeterminacy

Between Gift and Exchange

The above lines have touched on the second way in which elementary transfers can be made more complex: that of indeterminacy. In state societies, at least in principle, the boundary between what is legally enforceable and what is merely morally obligatory is clear, and the (written) law is supposed to draw it with precision. The rigour of this delimitation should certainly not be exaggerated: reality can deviate significantly from the principles, either because of the financial or social obstacles that can dissuade or even prevent the unworthy from filing a complaint, or because it is never possible to enclose reality completely in a few formal precepts — the activity of tax advisors, for example, consists to a large extent in taking advantage of these grey areas. But the indeterminacy of exigibility which, where the state reigns, can be interpreted as a deviation from the norm, represents a much more banal case in non-state societies, for an additional reason: within the framework of a private dispute, the implementation of violence, however legitimate, is to be borne by the injured party. In simple terms, if in a state society, one does not take justice into one's own hands, in a stateless society this is not only a legitimate means of asserting one's rights: it is the only possible one. In these conditions, it is easy to imagine that the act is conditioned by the balance of power. An individual without support, or a member of a group of relatives that is too weak, will be able to give up using violence that is otherwise socially legitimate, simply because he does not have the means to do so. The concrete manner in which legal obligation is exercised in stateless societies thus favours the existence of social situations in which exigibility is presented in an attenuated form, and all the more difficult to discern.

This issue is particularly frequent in ethnology with a form of bilateral transfer which, at first sight, belongs to both exchange and gift — it is precisely this case of figure which led Marcel Mauss, and many others after him, not to establish a clear conceptual differentiation between the two terms. According to Alain Testart, who devoted many pages to analysing it (Testart 2001), such a configuration must be understood as a non-market exchange, as opposed to the market exchange, which is canonical to our own society.

The market exchange is the one in which the sale of the good is independent of the personal link that unites the seller and the buyer. Ultimately, this link no longer exists: the market is an anonymous place, where anyone can buy anything from anyone — to the point that, once the goods are put up for sale, refusing to conclude the transaction in view of the identity of the buyer is nowadays a penal offense. On the other hand, a non-market exchange is an exchange framed, or even conditioned, by the prior personal link between the partners. It is the pre-eminent form of exchange in ancient societies, which ignore the market and are steeped in person-to-person ties and where, very often, exchange friendships are formalized and named, as among the Jivaro (Harner 1973, 125).

In remarkable pages, Alain Testart puts in comparison the two opposite forms of fetishism generated by these two categories of exchange. The fetishism proper to commercial exchange, underlined by Marxist economic analysis, makes the value of goods appear as their intrinsic property, while concealing the fact that this value expresses a social relationship between producers. Conversely, writes Alain Testart:

> If the market form generates its own illusion, the non-market form can also generate its own. That which is inserted in a relationship of friendship [...] engenders the illusion according to which it would only be a question of gifts and counter-gifts. This exchange takes on the appearance of an exchange of gifts [...]. No mention is made of a promise to return, as if not to offend the susceptibility of the 'friend'; even less is he reminded that he is in debt, if this is the case, for even more obvious reasons; the successive acts, which in reality answer each other and group themselves two by two as cessions and counter-cessions, appear as if they were not linked together; the very essence of the exchange, the fact that it is a conditional cession [...], and the exchange can henceforth appear only as a series of acts based on the sole freedom of each person and the good feelings in which he finds himself with regard to his partner. (Testart 2001, 739–40)

In a correlative manner, the dividing line between exchange and gift — the exigibility of the counterpart — tends to be faded and to become often more theoretical than real:

> Whatever the reason for non-return, betrayal by the friend, bad grace or simple material impossibility, it is only very rarely that one will have recourse to this ultimate means [violence]. One would have to mobilize one's relatives, all those who can and are willing, for one reason or another, to lend a hand; it is always a risky operation, it is a war. This is why one will generally be satisfied with breaking the relationship which existed between the two friends: one will not exchange any more and one will not have any more relationship with him. This is precisely what a donor does to an ungrateful recipient who does not provide a counter-gift. The usual mode of sanction of the exchange between friends is the same as in the gift. (Testart 2001, 741)

Alain Testart then adds in a note:

> This is also why it is so difficult to disentangle in ethnographic data what is gift and what is exchange between friends: the common mode of sanction is the same, and it is only the presence, or the possibility admitted as legitimate, of a rare mode (and all the more so since the societies studied are always in the colonial context), the violent mode, which in all rigor can provide the decisive criterion. (Testart 2001, n. 39)

The question that can nevertheless be asked — and which was not addressed with sufficient clarity in the previous text that I devoted to this subject — is that of the existence of an objective zone of indeterminacy. The problem goes beyond the simple 'difficulty', on the part of an outside observer with incomplete information, of ruling on the possible exigibility of a counter-transfer. It is also to know if, in the eyes of the parties concerned as well as of the society to which they belong, there can exist objectively indeterminate situations. As mentioned above, such situations, although rare, exist in societies with written law and a formalized judicial system. In the case of societies that do not have such systems, there is no reason to exclude them from the analysis.

Thus, alongside non-market exchange, which, while possessing the outward appearance of a gift, is indeed exchange, we must concede the existence of a 'circle-square', the 'gift-exchange' — for, if in the abstract realm of theory, social and geometric bodies are pure, in the real world, they can borrow their characteristics from two distinct types.

Would all this mean a return to the starting point? After having criticized them so severely in the wake of Alain Testart, are we not now recognizing the relevance of the concepts proposed by Mauss? Wouldn't the conceptual demarcation between exchange and gift be vain, or, at the very least, irrelevant for societies without a state or a market? Such a conclusion would be tantamount to throwing out the baby with the dirty bathwater. Not only can an intermediate situation only be properly understood if the two poles that constitute it have been properly defined, but the case under which Mauss, and others after him, grouped all transfers in stateless societies only constitutes a very specific category.

Other Indeterminacies

To conclude on this point, it should be added that the indeterminacy of exigibility does not only blur the boundary between gift and exchange, and that it can just as well concern the other modes of transfer.

Compulsory exchange remains a relatively rare form, so it is hardly surprising that its fringe is even rarer. In modern times, however, an intermediate form between free and compulsory exchange can be discerned in certain types of wage relations. Indeed, these have not always taken the form they do in the modern labour market. As Yann Moulier-Boutang (1998) has shown in his monumental work on this question, between slavery and free wage labour there have been countless situations where formal freedom was intermingled with legal obligations, in order to force labour to be exchanged (or to keep a freely initiated exchange going). This 'constrained wage system' took, for example, the form of peonage, widespread in Latin America, where the supposedly free worker was chained to his employer by a system of debts that were never extinguished.

Finally, exigibility may also suffer from indeterminacy without the transfer being conditioned by a counterpart. The most obvious case is that of an intermediate situation between the gift and the t3t, namely a unilateral transfer that would stand at the limit of exigibility. The medieval period provides various examples of transfers destined for an emerging or failing state, and of which it is not known whether they were taxes or gracious gratuities (Guéry 1984, 1256–59). It is necessary to repeat that exigibility exists, however, outside the state. Concerning societies studied by ethnology, the possibility of indeterminacy between t3t and gift is exemplified by the custom known among various hunter-gatherer peoples under the name of demand sharing. Nicholas Peterson (1993), the scholar who first named the phenomenon, groups together under this same name situations of a very different nature from the point of view of exigibility. The dimension that he emphasizes — the fact that the transfer is solicited by its recipient, whether verbally or not — does not in fact overlap with its obligatory nature. There can therefore be gifts that are solicited (Christmas presents to children, after a duly drawn-up wish list) or not (one made spontaneously to a friend in need), as well as obligatory transfers that are solicited (a fine paid after a parking ticket) or not (taxes paid by direct debit). In itself, therefore, the dimension of solicitation does not seem to be very useful for grasping the dimension of obligation. Nevertheless, it is clear from the examples given by Peterson that, while some are clearly gifts and others are compulsory transfers (such as the transfer of certain parts of a game animal as part of a kinship relationship), there are cases where solicitation exerts pressure bordering on a threat, without it being possible to determine whether this threat refers to a socially legitimate exercise of violence.

By the way, these indeterminacies do not only concern (legitimately) payable transfers. Anthropology has also drawn attention to the phenomenon known as tolerated theft (Blurton Jones 1987). While its study has focused on the issue of its social functions and its place in evolutionary reasoning, it is not without interest to note that this practice, by standing on the borderline between theft and gift, completes the zones of indeterminacy by providing an intermediate form between gift and what could be called 'transfers of the fourth type' (t4t), i.e. illegitimate transfers (Athané 2011).

From Forms of Transfer to Social Relationships

In what way does this social analysis, which could be described as 'grammatical', articulate with the archaeological approach?

At the risk of giving the impression of sawing off the branch on which this text is sitting, it is necessary first of all to insist on the innumerable — and sometimes insoluble — difficulties that this connection raises, and which fall into two different categories. The first are internal to sociological theory, and have to do with the complexity of the relationship that modes of transfer have with social structures. The second are specifically archaeological, and are linked to the fact that this discipline can only apprehend these structures and social relationships through material facts.

Forms of Transfer and Social Relationships

As we have just said, there is no simple and univocal relationship between the formal nature of a transfer and the social relationship in which it is embedded. More precisely, while the formal nature of a transfer is an indispensable element in the characterization of the social relationship that involves it, it is far from sufficient.

To be convinced of this, let us consider the two unilateral transfers represented by the gift and the t3t, which are radically opposed in terms of exigibility. Contrary to what one might believe at first sight, this opposition does not correspond in any way to the absence or presence of a hierarchical relationship. If, in our societies, the exigibility of the t3t immediately brings to mind transfers imposed by the authority of the state, this exigibility can intervene in quite other contexts, in particular that of kinship. This situation is obviously very common in non-state societies. But even in contemporary society, let us recall that parents have an obligation of maintenance towards their children, which obviously does not express the slightest hierarchical superiority of the latter over them. In the same way, giving is very far from always being the expression of a disinterestedness and an absence of domination. It can be made by the superior to the inferior, on an individual basis (the alms given to the poor, the liberality to the Roman client) or collectively (the ancient evergetism or our modern philanthropic foundations). In all cases, the rich person honours himself by publicly accomplishing his gesture and by ensuring that it has all the desired resonance. Conversely, the gift can also emanate from the subordinate who solicits the powerful, including in the form of sacrifice when the latter belongs to the supernatural world.

Moreover, if we place ourselves this time not at the level of a particular social relationship, but at the more general level of the social type, it would be futile to hope to characterize it by the absence or presence of a given form of transfer; the four elementary forms of transfer probably coexist in all social types. On the other hand, some types may be marked by the predominance of certain forms over others. In a society such as that of the San ('Bushmen'), the vast majority of transfers were in the form of gifts — or some indeterminate form between gifts and t3t. Capitalism, this 'immense collection of commodities' (Marx [1867] 1990, 125), is dominated by exchange — although the strong tendency to an increasing role of the state in the economy gives ever greater importance to t3t.

However, this approach remains a clumsy way of posing the problems. The usefulness — or rather, the indispensability — of a robust classification of modes of transfer is not that it can be used to determine a social relationship by itself. It is to contribute to this determination, in particular by making it possible to discriminate between two social relationships that could wrongly be confused, and thus avoiding Mauss's mistake of thinking of all 'primitive' transfers in a single, vast category. This is certainly of more modest scope, but it should not be underestimated. Thus, for example, the criterion of exigibility makes it possible to differentiate between gifts made within the framework of a client relationship and tributes and other contributions imposed by a chiefdom or a state. The same is true, for example, of vast collective achievements, where one must ask whether the work was provided freely or whether it induced coercion — the social relationship between the ordering party and the performers being obviously not at all the same in one case as in the other. This last point once provided the material for lively discussions about, among other things, the pyramids of Egypt or, more recently, the megaliths (Boulestin 2016; Gallay 2006; Testart 2014).

Archaeology and Forms of Transfer

To the internal difficulties in the articulation between types of transfer and a theory of social relationships and types, one must add those which are specific to archaeology, and which are far from minor.

To begin with, it is necessary to recall two trivialities, which nobody denies on principle, but which are too often lost sight of when interpreting the data.

The first, already recalled in the introduction to this text, is that unless one has an inscription clearly playing the role of a property mark, it is extremely difficult, if not strictly impossible, to know the identity of the holder of the rights exercised over an object found in an archaeological context, and not even the nature of these rights. At most, one can make more or less well-founded suppositions, based on indirect elements (such as the habit of attributing to the deceased the goods found in his grave). It should be added that if, in the absence of a written record, one cannot know who held the rights to a property, one cannot consequently know either if these rights had changed hands, and under what conditions. In other words, ownership is entirely subject to archaeological invisibility — and the nature of the transfers to which it gave rise with it.

The only dimension that archaeology apprehends with more certainty — it therefore speaks of it all the more willingly — is the spatial one. It is sometimes possible to demonstrate that an object found in a given place comes from somewhere else. Some movements are spectacular, such as those of shells found far from the coast, the extraordinary 'laurel leaves' of Volgu, carved twenty thousand years ago in blocks of flint extracted from veins located up to 250 km away (Thevenot 2019, 61), or the polished alpine jade blades emblematic of the Linear pottery culture (LBK), which are found up to 2000 km from their place of production (Gauthier and Pétrequin 2017). This is commonly referred to as the 'circulation' of these raw materials, or so-called prestige goods. But this term circulation is as misleading as it is useful. The only circulation that archaeology can detect is physical circulation; and this physical circulation is very different from economic (or social) circulation, i.e. the transfer of property or usage rights. The first indicates the probability of the second: it is difficult to imagine how farmers or hunter-gatherers would have regularly travelled hundreds of kilometres to obtain supplies of flint, for example. Still, one must be wary of hasty conclusions. In Irian Jaya, the Dani studied by Pierre and Anne-Marie Pétrequin (2002, 134–46) travelled several days to get to the quarries from which they got the stones for their polished axes. And in Australia, groups could travel several hundred kilometres to acquire ochre (Jones 1984). In any case, if spatial circulation over long distances makes it possible to deduce with reasonable certainty social circulation, in the opposite direction, a good part of the social circulation does not translate into any significant physical circulation and therefore remains archaeologically invisible: to take the case of the Dani, if the axe is used to pay a bride price to a family of the same village, this social transfer will only translate into a physical movement of a few metres. In other words, archaeology will sometimes be tempted to see a social circulation where there was none, and will often have no way of perceiving it where it existed.

Conclusion

The rigorous classification of the modes of transfer of goods constitutes an essential element for the theory of social relationships and, beyond, of social formations. It will only be possible to go beyond the rather heterogeneous and ill-defined categories in use ('bands', 'trans-egalitarian societies', 'mid-range societies', 'chiefdoms', etc.) by endeavouring, among other things, to forge a new classification based on a finer characterization of social relationships, within which the transfer of goods most often plays a key role. It is only in this way that social anthropology, reviving the scientific programme around which it was once constituted, will be able to provide the conceptual tools that will make it possible to link material production and social structures, and thus provide archaeologists with the keys to interpretation (Testart 2006). For the latter, on the other hand, and because of the very way in which their discipline approaches social facts, the nature of the transfers to which the goods they discover gave rise — and, most of the time, the very existence of these transfers — is not a key, but a closed door. Archaeology has no direct way of establishing whether the jade axe it uncovers was given away, sold, or given up as a mandatory payment, and any attempt to rule on this question directly from the characteristics of the object can only be a speculation. The answer, however, may be reached by more circuitous routes. By relying on a solid anthropological theory — which has yet to be built — the archaeological investigation can try to determine the type of society that produced this axe and thus, indirectly, identify the social relationships in which it was inserted as well as, consequently, the probable manner and circumstances in which it was brought to change hands.

Works Cited

Athané, F. 2011. *Pour une histoire naturelle du don* [A Natural History of the Gift] (Paris: Presses universitaires de France).

Blurton Jones, N. G. 1987. 'Tolerated Theft, Suggestions about the Ecology and Evolution of Sharing, Hoarding and Scrounging', *Social Science Information*, 26.1: 31–54 <https://doi.org/10.1177/053901887026001002>.

Boulestin, B. 2016. *Mégalithismes vivants et passés: approches croisées* [Past and Present Megalithisms: Crossed Approaches], ed. by Christian Jeunesse, Pierre Le Roux, and Bruno Boulestin (Oxford: Archaeopress).

Darmangeat, C. 2016. 'Don, échange et autres transferts' [Gift, Exchange, and Other Transfers], *L'Homme*, 217.1: 21–43.

Feil, D. K. 1980. 'Symmetry and Complementarity: Patterns of Competition and Exchange in the Enga Tee', *Oceania*, 51.1: 20–39 <https://doi.org/10.1002/j.1834-4461.1980.tb01417.x>.

———. 1984. *Ways of Exchange: The Enga Tee of Papua New Guinea* (St Lucia: University of Queensland Press).

Gallay, A. 2006. *Les sociétés mégalithiques: pouvoir des hommes, mémoire des morts* [Megalithic Societies: Power of Men, Memory of the Dead] (Lausanne: Presses polytechniques et universitaires romandes).

Gauthier, E., and P. Pétrequin. 2017. 'Interprétations sociales des transferts de grandes lames polies en jades alpins dans l'Europe néolithique' [Social Interpretations of Transfers of Large Alpine Jade Polished Blades in Neolithic Europe], *ArcheoSciences: revue d'archéométrie*, 41.1: 7–23 <https://doi.org/10.4000/archeosciences.4856>.

Guéry, A. 1984. 'Le roi dépensier. Le don, la contrainte et l'origine du système financier de la monarchie française d'Ancien Régime' [The Spendthrift King. Giving, Coercion and the Origins of the Financial System in the French Monarchy of the Ancient Regime], *Annales*, 39.6: 1241–69 <https://doi.org/10.3406/ahess.1984.283129>.

Harner, M. 1973. *The Jívaro: People of the Sacred Waterfalls* (London: Hale).

Jones, P. 1984. 'Red Ochre Expeditions: An Ethnographic and Historical Analysis of an Aboriginal Trade in the Lake Eyre Basin. Part I', *Journal of Anthropological Society of South Australia*, 22.7: 2–10.

Lederman, R. 1986. *What Gifts Engender: Social Relations and Politics in Mendi, Highland Papua New Guinea* (Cambridge: Cambridge University Press) <https://doi.org/10.1017/CBO9780511753022>.

Marx, K. [1867] 1990. *Capital: A Critique of Political Economy*, Penguin Classics (London: Penguin).

Mauss, M. [1924] 2004. *The Gift: The Form and Reason for Exchange in Archaic Societies*, repr. Routledge Classics (London: Routledge).

Moulier Boutang, Y. 1998. *De l'esclavage au salariat: économie historique du salariat bridé* [From Slavery to Wage Labour: The Historical Economy of Bridled Wage Labour] (Paris: Presses universitaires de France).

Peterson, N. 1993. 'Demand Sharing: Reciprocity and the Pressure for Generosity among Foragers', *American Anthropologist*, 95.4: 860–74.

Pétrequin, P., and A. M. Pétrequin. 2002. *Ecologie d'un outil: la hache de pierre en Irian Jaya (Indonésie)* [Ecology of a Tool: The Stone Axe in Irian Jaya (Indonesia)] (Paris: Centre national de la recherche scientifique).

Testart, A. 2001. 'Échange marchand, échange non marchand' [Market Exchange, Non-Market Exchange], *Revue française de sociologie*, 42.4: 719–48 <https://doi.org/10.2307/3322738>.

———. 2006. 'Comment concevoir une collaboration entre anthropologie sociale et archéologie? À quel prix? Et pourquoi?' [How Can We Make Social Anthropology and Archaeology Work Together? At What Cost? And Why?], *Bulletin de la Société préhistorique française*, 103.2: 385–95.

———. 2007. *Critique du don: études sur la circulation non marchande* [Critique of the Gift: Studies on Non-market Circulation] (Paris: Syllepse).

———. 2014. 'Anthropology of the Megalith-Erecting Societies', in M. Besse (ed.), *Around the Petit-Chasseur Site in Sion (Valais, Switzerland) and New Approaches to the Bell Beaker Culture* (Oxford: Archaeopress), pp. 331–36.

Thevenot, J.-P. (dir.). 2019. *Les silex solutréens de Volgu (Rigny-sur-Arroux, Saône-et-Loire, France): un sommet dans l'art de la 'pierre taillée'* [Solutrean Flints from Volgu (Rigny-sur-Arroux, Saône-et-Loire, France): A Summit in the Art of Cut Stone] (Dijon: Revue archéologique de l'Est).

ALEKSANDER DZBYŃSKI

2. Between East and West

The Measurement of Linearity as the Forerunner of Weighing

ABSTRACT A number of artefacts made of metal in Europe before the Bronze Age unambiguously show that they were measured and perceived numerically. They are usually described as beads, wires, and sheets of metal. A number of tokens, made of clay, have also been found in southern and eastern Europe in the Neolithic but, unfortunately, we still don't have an appropriate idea of how to interpret them. Are they to be addressed similarly to their Near Eastern counterparts? Or has their role changed after arriving in Europe and the following adaptation into local environments and cultures? Moreover, there is an additional group of evidence and indices showing that the development of numbers and measures was a continuing process not only in the Near East, but also in the Neolithic and Eneolithic in Europe. In this paper, I would like to present a summarizing attempt to interpret the main development paths of Europe and the Near East. On the one side, we have the Near Eastern phase of number and measure development constituted by artefacts such as tokens, that led in the subsequent phase to the establishment of full numeracy, literacy, and centralized power. In Europe, however, we see a strong current of early metal production that introduces numbers and measures, as it seems, by another means of cognitive conceptualization. The interpretations presented open up a series of questions concerning historical, cultural, and cognitive processes on both sides of the Old World.

KEYWORDS Linear measurement, weight, prehistory, Near East, Europe, real praxis, ideal praxis, portions

The East

The Neolithic period witnessed many changes in terms of society and economy. During this time we observe the existence of permanent settlements in the archaeological record, population growth, enrichment of material culture, and, among many other things, 'memory overload'. By this I mean the cognitive capacity

Aleksander Dzbyński University of Zurich and Warsaw University. Aleksander Dzbyński studied prehistoric archaeology, anthropology, and ancient art history at universities in Warsaw, Basel, and Freiburg, and he received his PhD degree from the University of Basel in 2004. He has worked as a Senior Lecturer in Archaeology at the University of Rzeszów and at the University of Zürich. Currently, he is working as a Privatdozent at the University of Zürich and a full time Associate Professor at Warsaw University, where he teaches prehistoric archaeology and archaeological methods and theories. His methodological approach builds upon archaeological theory, anthropology, and philosophy, whereas his area of specialization concerns material sources of mathematization of the human mind in prehistory which has been the theme of his three previous books: *Metrologische Strukturen in der Kultur mit Schnurkeramik und ihre Bedeutung für die Kulturentwicklung des mitteleuropaischen Raumes* (Beier & Beran, 2004), *Ritual and Understanding: Rational Bases of Communication and Exchange in Prehistoric Europe* (University of Rzeszów Press, 2008), and *The Power of the Line: Metaphor, Number and Material Culture in European Prehistory* (Cambridge Scholars, 2013). He also wrote a non-fiction book: *Mr Blademan and the World-Machine: Man vs. Technology – from the Stone to the Computer* (Sorus, 2010) and a theory-grounded book *Cinematosophical Introduction to the Theory of Archaeology: Understanding Archaeology through Cinema, Philosophy, Literature and Some Incongruous Extremes* (Vernon Press, 2020).

Number and Measure in Prehistory: Gift, Exchange, and the Transfer of Goods, ed. by Philippe Della Casa and Aleksander Dzbyński, WSA, 6 (Turnhout, 2024), pp. 25–40.
BREPOLS PUBLISHERS
DOI 10.1484/M.WSA-EB.5.137127

of humans to recognize members of their own group that had evolved in a simple social environment. This cognitive capacity is sometime called 'Dunbar's number' because it was first proposed in the 1990s by the anthropologist Robin Dunbar, who found a correlation between primate brain size and an average social group size (Dunbar 1992). By using the average human brain size and extrapolations from the results of research on primates, he proposed that humans can comfortably maintain 150 stable relationships. Dunbar (1998) explained it informally as 'the number of people you would not feel embarrassed about joining uninvited for a drink if you happened to bump into them in a bar'. In Neolithic settlements of the Near East, the number of inhabitants increased up to a few thousand people. Brian Hayden (1995) estimated for example a 1600 per cent increase in human population from the Natufian to the PPNB period in the Levant (*c.* 10,000 to 7500 BC). In such an environment, everyone wants to benefit: to communicate, to exchange things, etc. This is a communication problem that comes with the Neolithic. For the first time in history so many people can share, exchange things, news, and services living enclosed in a small space for the most part of the year. It seems that doing this with the help of old methods of storytelling and gift giving alone would not have sufficed. There was a need to develop new ways of communication.

The solution to the problem were recording systems. Small geometric objects, which according to Material Engagement Theory (MET) were not numbers yet, but enactive signs, something like material harbingers of numbers. Lambros Malafouris (2013) believes that in these early stages of development, both on an ontogenetical and phylogenetical level, the use of tokens provided the necessary external scaffolding for the emergence of arithmetic competence. The process of token counting should be seen as an 'integrative projection' between the basic biological 'numerical sense' and the physical fields of experience (body parts, fingers, tokens). It is the resulting structural coupling or mixture that gives the possibility of significant cognitive action, which we call counting, and not just some innate biological capacity of the human brain. Clay tokens do not mean, as it may seem in common sense, numbers. They extract numbers and make a visible and tangible manipulation of their properties. What essentially happens in this case — put in very simple terms — is that the vague structure of a very difficult and inherently meaningless conceptual problem — that is, counting — by being integrated via projection with the stable material structure of the clay tokens, is transformed into an easier 'spatio-visual problem'. However, spatio-visual problems can be directly manipulated and manually resolved in real time and space. According to Schmandt-Besserat (2010) the tokens made it possible to visualize and manipulate numerocity. Thus the problem — that is, counting — becomes meaningful (Malafouris 2010, 40). The meaning of number comes therefore from the external world, and its source is basically material culture which becomes much more easy to control and shape.

As mentioned, the connection between this material solution to the evolutionary dependent cognitive problem is evidenced in the archaeological record in the Near East. It has been researched in detail at the site of Ain Ghazal in the Jordan Valley. Here, there is evidence that Ain Ghazal grew up from 5 to 10 ha during the PPNB, to reach a massive 13 ha during the PPNC at around 6750 BC. The site was 'briefly booming', as Bellingwood (2005, 65) phrased the circumstances in question. Large numbers of tokens were used in Ain Ghazal at that time.

The famous media theorist Marshall McLuhan (1964) believed that the number is an extension of the sense of touch. His thesis can be also confirmed in the Near East. Tokens were material metaphors — they represent something in the outside world with which this world can be both physically and mentally manipulated. Thus, taking into account the above findings, we can distinguish several factors that tokens introduce into the Neolithic culture.

According to MET, they introduce a material anchoring and embodied experience (Malafouris 2010; 2013) which allows new senses to develop in human communication. Second, we can put forward the hypothesis that their use was related to the formation of the collection metaphor, which is one of the basic grounding metaphors according to the concept of George Lakoff and Rafael Núñez (2001). In addition to this, Karenleigh Overmann's findings should be pointed out. She stressed that in the Neolithic a new mathematical feature arrived that was the 'bundling' (Overmann 2018). Initially, it was Schmandt-Besserat (1992; 2007) who suggested this by claiming that larger tokens of the same shape meant groups of objects, such as a herd of animals. In a nutshell, tokens are the first objects that make the grouping of elements visible, which allows us to assume the existence of a counting base. Finally, tokens were also something like tokens of debt (Graeber 2011). This assumption is logical and stems directly from the original concept of Schmandt-Besserat (1992), who defined them as recording systems. They recorded mutual obligations between people, usually between an institution (a temple or a palace) and people. And last but not least, they meant something more or, better to say, something less than a message. The entry that was placed in the register meant a reduction of message because the emphasis was on measure and number, not on descriptive, narrative quality, which had previously been the norm in exchange relationships (Dzbyński 2008a). So here we would be dealing with the mediatization of communication according to Habermas's concept, as well as with its rationalization (Habermas 1981).

In this context it will be useful to outline the differences between the earlier systems of Palaeolithic notation and tokens. The first difference is that Palaeolithic notation systems are one-dimensional and tokens are two-dimensional. This difference is emphasized by Overmann. In one-dimensional notations, accumulation can be recorded but they do not have grouping potential. Potential relations between numbers are those of an ordinal counting sequence: more than, less than, same as, etc. These intra-exponential relations are implicitly quantificational but not necessarily explicitly numerical, as Overmann states, giving the example of a rosary (Overmann 2018). Total value in this system is achieved by accumulating along the single dimension. Two-dimensional devices with grouping and manipulability potentialize the explication of 'inter-exponential relations', as well as operations beyond accumulation. They include the Mesopotamian clay tokens. Overmann goes on to suggest that in later periods tokens may have been used with counting boards, making them an abacus-like device.

What happens later in the prehistory of the Near East seems to be a consequence of the phenomena defined above. It is particularly interesting in this context that in the idea of measuring in the Near East grain takes the prominent role. The wages are paid in grain and grain is brought to the temple as a sacrifice or tax. Grain is at the heart of the interests of society at the time. Grain was so important that it was probably used as a source of metrology. This must have been an ingenious invention. Both weight and length were considered to be a certain count of grain kernels and were set up to different metrological units. In this case we could say that metrological systems in the Near East were actually based on counting, on putting all the small grain kernels together and translating them into different measurement units. Even the shape of the first weights takes the form of a single grain kernel. In terms of mathematical cognition this refers to the collection metaphor proposed by Lakoff and Núñez (2001).

The Neolithic culture arrived in Europe from the Near East. Its occurrence is initially limited, however, to south-east Europe — an area culturally linked to the direct colonization of Anatolia, or to subsequent influences from this direction. Both the material and spiritual culture of this region still clearly resemble the Anatolian one, although it is certain that the newcomers had to react dynamically each time to changing environmental conditions, they had to experiment with new crops or animals, so the spread of agriculture could not have taken place mechanically. As a result, the Neolithic of temperate Europe began to differ significantly from that of Greece or the Middle East.

The history of research on the European tokens resembles what had been proposed for the Near East. Originally, they were seen as objects with unknown functions, defined by more enigmatic terms such as clay cones, miniature clay objects, buttons, pawns, etc. In the earliest phases of the Neolithic, they are not numerous. Only a dozen or so pieces were found in the Protosesklo phase on sites such as Argissa, Sesklo, and Pyrasos. They appear a little later, e.g. in layers dated 6800–5800 BC in Thessaly, where they can be identified with the Sesklo culture. In this phase, settlements develop, houses become larger, there are even megaron-shaped buildings with two or three rooms. The size of the settlements increases to 2 ha. At first glance, it seems that the appearance of tokens in south-east Europe was connected to the spread of the Neolithic. However, they did not appear in the initial phase of Neolithic culture, in the phase of experimentation in the new environment, but only in the phase of consolidation and flowering. Perhaps at that time contacts with the homeland of agriculture intensified and changed their character? According to Mihael Budja (2003), a disproportionately large number of tokens were found in Nea Nikomedea (*c.* 6000 BC), associated with white and red painted ceramics. This site is seen as a transmitter of cultural currents from Anatolia.

Typological analogies of European tokens with their Near Eastern counterparts also pose some problems. In Europe, the early stage of simple and complex tokens cannot be distinguished. On the contrary, tokens of this type appear much later in Europe than in the Near East. Nevertheless, researchers are inclined to assign small artefacts found at sites such as Argissa, Souphli Magula, Achilleion, Sesklo, Genticoi, and Vrbica to vessel-shaped tokens, i.e. the equivalents of tokens from the Middle East. These would therefore be much older than their Near Eastern counterparts and by this they are of particular importance according to Budja (2003). They occurred, among others, at sites in Dalmatia and in Croatia, connected with Impresso ceramics. They were also found in graves, which, in his opinion, is also interesting because of the parallels with the Near East. In the Near East, tokens were usually found in luxurious or otherwise distinguished graves (Schmandt-Besserat 2007). Despite many difficulties in interpreting these mysterious objects, Budja has been able to demonstrate how the south European tokens, in addition to the already significant formal differences from their Middle Eastern counterparts, function in a network of interregional communication in south-east Europe. He also proposes to see in them

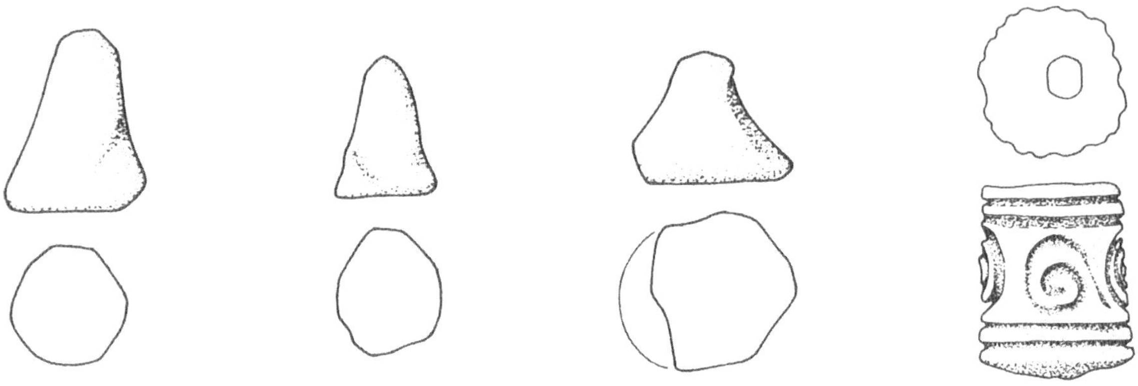

Figure 2.1. European tokens (after Budja 1998).

the manifestations of highly individualized and distinct dynamics of cultural change in this part of the continent. According to him, different types of tokens were developing here, which can hardly be compared with the Near Eastern specimens. Unfortunately, there is no systematic research on European tokens. Budja goes boldly in his interpretations, suggesting reinterpretation of many objects. He suggests, for example, that the hitherto 'zoomorphic amulets' should also be treated as tokens, as a result of contacts with local hunting communities which co-shaped the cultural image in Europe. The basic problem of European tokens would therefore be a significant formal distinctiveness from their Near Eastern counterparts. Moreover, they are not high in number when compared with the Neolithic initial area (Fig. 2.1).

The context of European findings differs from the Near Eastern original. Above all, in Europe a redistributive system that eventually dominated Near Eastern communities, determining their trajectory of development to theocratic states, has not developed. This model was defined as staple finance, while the European trajectory was characterized by Kristiansen as wealth finance (Kristiansen 1991), due to lack of centralization of power in larger areas, and due to focus of social attention on luxury goods. Roughly speaking, this type of social model can be described as a chiefdom.

Budja points to three examples (Nea Nikomedea, Rakitovo, Donja Branjevina) in which European tokens appeared in almost identical sets: anthropomorphic figures, clay altars, and zoomorphic vessels. It seems, therefore, that European tokens have been more strongly associated with ritual functions. This finding points to Chapman's work on the phenomenon of fragmentation of objects in southern Europe, where the author writes that the fragmentation of anthropomorphic figures was connected with the practice of enchaining human relations (including the enchaining of relations with ancestors) by material means — 'the tokens of significance' — which included anthropomorphic figures and their fragments (Chapman 2000, 75). In this perspective, therefore, tokens in European contexts seem to have a broader, less specialized meaning than in the Near East. Perhaps they were more tokens of significance than tokens of number. Moreover, the south European tokens seem to have a similar epistemological status as the so-called 'Balkan writing'. In the Balkans, both tokens and 'writing' can be recognized as a distinct element of religious rituals. In the Near East, however, it is a rationalized administration and trade. Tokens are also found at sites of the Tripolye culture in eastern Europe, covering more or less today's western Ukraine, but there is a lack of systematic studies so far. The issue of European tokens is still open for research and discussion.

The West

Wealth finance, as defined by Kristiansen (1991), becomes evident in European cultures and appears to be based on metal — the early copper industry — quite early on. A renowned example is the cemetery in Varna by the Black See Coast (Bulgaria). We can observe here for the first time a group of very rich graves, equipped with a great amount of metal (copper and gold) — the first elites of those time (*c.* 4500 BC). The combination of gold and copper indicates that items without any utilitarian value have been deposited in Varna's graves even though formally they are tools. In the richest grave, no. 43, to which we will return, almost one thousand gold objects weighing 1.5 kg were deposited. Today's value is about 100,000 USD.

My question has always been: How did the people of Varna measure this extraordinary richness of metal? For example, Jan Lichardus (1991) believed that there had to be some kind of established weighing system in Varna, especially with regard to gold. This thesis, although it may seem intuitively right and commonsensical, may not be valid for the following reasons. First of all, we have no such evidence of any kind, neither material nor indirect, similar to the existence of adequate solutions in the Near East at that time. The present sources so far clearly indicate that weighing systems developed quite late, namely in the Early Bronze Age civilizations in the Middle East and Egypt in the first half of the third millennium BC (Rahmstorf 2010; Rahmstorf in this volume). These systems were established almost two thousand years after the Varna culture, where neither urban structures nor central authorities developed. The Varna culture itself quickly disappeared and its development lasted only two to three hundred years (Chapman and others 2006). Therefore, we should either assume that in the Varna culture a weight measurement system had been invented much earlier than in the Near Eastern civilization, and was then quickly forgotten, or that metal was measured in a completely different way at that time.

Measuring weight is, contrary to its modern appearance, a quite complicated process. It requires an enforcement of significant cognitive work involving the formulation of ideal mathematical concepts, i.e. abstract numbers. This did not suddenly appear in human culture, so there had to be some stage of formation, a process of arriving at this point. The research on this issue conducted in the Near East suggests that it was a long-term process (Damerow and others 1988). Admittedly, research on early mathematical perceptions has essentially been reduced to numbers and counting, so there is a significant gap in our understanding of early metrology. However, based on theoretical considerations on the one hand, and on what we know about numbers on the other hand, we can conclude that it must have been a process that went from a concrete measure which was apparent — from its visibility towards ever greater abstraction. Here we must first of all realize that the weight is an abstract measure, separated from the visibility of the anthropomorphic linear measure, and that the weight balance itself is a certain device (lever) measuring without human intervention. So it is quite a complicated device, which requires equally complicated theoretical knowledge, and theoretical knowledge is created only where there are conditions for its development. From our point of view, it is enough to find a point of equilibrium for a properly prepared rod or stick and thus compare the weight of objects. But to do this one needs to be aware that the two sections of the stick are perfectly equal (the idealization of the linear section and its division into geometrically equal sections must be completed on the cognitive level) and a balance scale, supported by the existence of precisely balanced weights, representing abstract units, must be constructed which is again a complex technical device. All evidence of balance scales comes from the third millennium BC, from the environment of early urban, literate civilizations. Perhaps first experiments were carried out to compare weights in order to prepare and exchange so-called weight standards, mainly metal bars, before more complex fractions and weight scales were used (Michailidou 2010).

However, we have neither weights, nor bars made in the same size, nor any indirect evidence for using the concept of weight in Varna. Neither here, nor in the dozens of other cultures identified in the archaeological literature so far, has any such evidence been found before the Bronze Age. It should be noted that, technically speaking, the fifth millennium BC in Europe still means the Neolithic way of life. Most of the tools were still made of stone and flint, and actually all metal objects at that time still imitate stone or other previously known materials. For example, the ornamental objects in Varna often imitate shells, whereas gold beads imitate their earlier stone counterparts. A symbolic combination of old and new traditions is, for example, gold applications on Spondylus shells, from which spectacular decorations were made since the beginning of the Neolithic period (Sefériadès 2000). They were very important because they were exchanged almost all over Europe. Weight is an intellectual invention that makes it possible to rationally control the technology of metal production and its dissemination because it greatly facilitates calculations, exchanges, and transactions on a large scale. This is exactly what we observe in the Bronze Age, when the image of society changes completely: its structure, ideology, and economy (Kristiansen and Larsson 2005). But we cannot apply the same perspective to the Copper Age. The only problem probably is that we cannot imagine other ways of exchanging and assessing the value of metal than through its weight. But perhaps there were other ways?

A quick review of the ethnological literature allows us to imagine such ways. One of the best-known examples of using a simple linear measure to assess the amount of metal are the indigenous peoples of the North-West Coast of North America. These tribes used considerable amounts of copper and shaped it into the so-called 'coppers' — large sheets of copper, taking on a characteristic trapezoidal and ribbed form. It is possible that pre-contact societies made coppers out of native ores, but none of these survived. The real impetus for the cultural development of 'coppers' was the appearance of European trading ships on the North-West Coast in the 1790s (Macnaughtan 2020). Sheet copper soon became a prized trade item for all the coastal peoples from Alaska to Vancouver Island. Coppers were very valuable and meant wealth and prestige. They were only used in ceremonial transactions and potlatches — the most important ceremonies among the tribes.

A 'copper' might also be placed on the gravehouse or mortuary column to honour the dead (Emmons 1991). Although the entanglement of these extremely valuable copper products in socio-economic relations was very complex, evaluating the objects themselves was very simple. Coppers were divided into large, medium, and small ones. They were made in three standard sizes, according to the linear, anthropomorphic measure: the smallest baby 'copper' was about a hand length, the second was the length of the forearm to the elbow, and the third from the fingertips of the outstretched hand to the averted chin. The last size was worth as much as eight or ten slaves (Jopling 1989).

Coppers of the indigenous people of the North-West Coast allow us to imagine that metal can be evaluated in a different way than by weight, i.e. by an apparent, spatio-visual concrete measurement. At this point, I would like to point out artefacts that we should take into account in our discussion — the so-called flint long blades, superblades, or macrolithic blades — a quite mysterious technology that was developing at the time of the Varna culture and later.

The symbolic meaning of these blades is manifested in a way that seems obvious to everyone in the context of the Varna cemetery. Rich burials at the cemetery contained not only, as befits the era of copper, numerous objects made of this metal, but also one or more of these flint superblades. On the other hand, only fragments of such blades were found in less-equipped graves. We can therefore certainly assume that they were an important status symbol (like copper), meaning wealth and prestige. The researcher on the phenomenon of superblades, Laurence Manolakakis (2005, 300), writes:

> The maximum length of long blades increases with the richness of the tomb, even though shorter blades are present in all the tomb classes. The presence of longer blades, along with indirect indications such as fragments of different blades lined up to imitate the whole blade, show the value accorded to the notion of length in the funerary ritual of the Late Eneolithic and within the society.

Such blades also were a new communication medium. Researchers agree that they meant far-reaching contacts, prestige, new ways of symbolizing social roles as well as social differences. Such blades were clearly associated with wealth, prestige, and power. We should also not forget that we are talking here about a new technique of their production, which was known only to specialists who usually lived and worked in separate settlements. Specialists made concrete profits from their activities. In other words, the exchange of these items was based on power, economics, efficiency (specialized production), and a proper evaluation of these blades. A long time ago, the Polish archaeologist Stefan Krukowski (1922) called these blades the 'trade-blades'. They were produced in specific locations and then moved over long distances for exchange. They were also hoarded. The hoards of artefacts in general represent a special category of find, indicating particular attention paid to these artefacts, based on their real or anticipated values, or on their anticipated future use. In any event, the hoards suggest they were considered precious by their owners.

Based on Krukowski's thesis, we can basically say that we are dealing here with a forerunner of a mint. Such blades can be measured, hoarded, and disposed of, and they could constitute means of exchange (as opposed to the object of exchange). They can only be obtained from a few clearly defined sources of raw material, just like metal. They are produced by specialists who keep a permanent link to local structures of power. And unlike stone or flint axes, they have another important feature: their production is the most efficient of all — in comparison to the production of axes, which also were a valuable exchange item, but not as easily transferable and exchangeable as blades that could be immediately fragmented.

At this point, it is important to remember that these superblades, which appear as if on the margin of the metal-rich Varna burials, seem to be a necessary complement to a sumptuous burial, given that they were quite a concrete representation of the measure of wealth, at a time when it was not yet possible to put all the wealth of metal into abstract numbers as we do today through weight measurement. Fragments of blades were placed in poor graves, while the longest blades were deposited in rich graves (Fig. 2.2). The superblades were therefore a visible, apparent, and tangible expression of wealth that could not yet be fully rationally measured (i.e. in terms of weight). Symptomatic, in one of the richest graves in Varna, containing 225 golden objects, the longest blade, measuring 43.3 cm, was found. So we can say that the blades were as long as all the metal wealth deposited in the graves. Although they were the product of advanced craftsmanship (similarly to metallurgy), they still manifested the presence of archaic traditions, in the clash with the new phenomenon of accumulation, which could not yet be perceived in numbers, or did not need to be rationally measured (i.e. weighed). Varna is thus an illustration of the fundamental confrontation of the old world of narrative values (essentially the Palaeolithic world) with the world of accumulative values, which will become the foundation of the Bronze Age civilization. Therefore, the superblades could play a visualizing auxiliary and intermediary role in this process. But the most important conclusion from this is that it was about the length of these artefacts.

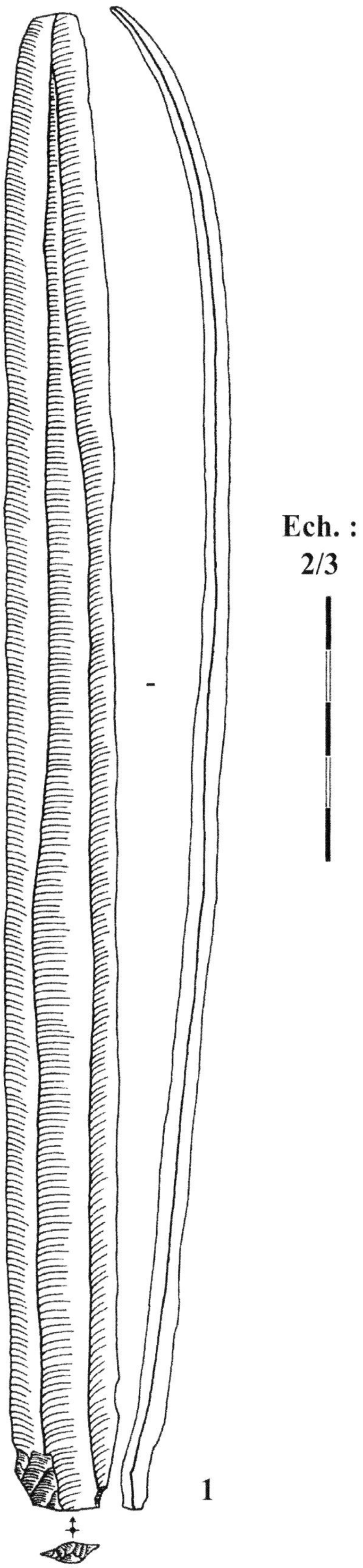

Figure 2.2. A superblade from Varna (after Manolakakis 2005).

In slightly later times the production of macrolithic blades was taken up, alongside intense metallurgical production, by communities inhabiting the Carpathian Valley and territories further north. The best-known and at the same time largest cemeteries from this period include Tiszapolgár-Basatanya (Bognar-Kutzian 1963) in Hungary and Tibava in eastern Slovakia (Šiška 1964). The burial rites in these cemeteries were clearly dichotomous, reflecting the sex of the deceased. The male burials contain characteristic objects such as copper and lithic adzes as well as flint axes (Lichter 2004). Copper and gold ornaments are to be found in male and female burials alike. Macrolithic flint tools were also characteristic of Tiszapolgár male burials in these cemeteries. Around one in five of these finds, made from imported Volhynian flint, are relatively large specimens, exceeding 10 cm in length, but most of the macrolithic finds recovered from burial sites had been severely fragmented before being deposited in the graves. The most important thing to note is the distinct correlation between the number of blades in inhumations of that period with the abundance of other grave goods or, in other words, with the relative richness of the graves' furnishings (Dzbyński 2011). While the rich graves yielded much greater numbers of blades, these were fragments rather than whole specimens. If we consider blade length distributions in rich and poor graves, the picture of blade fragmentation in the latter is not that clear. The blades in those graves represent the medium-sized categories, 5–8 and 10–14 cm in length. The corresponding situation in rich graves is more clear-cut. These yielded very short fragments, 1–4 cm in length, medium-sized specimens 7–10 cm long, and long macrolithic blades measuring 16–22 cm. These observations suggest that the metric characteristics of blades were subject to a specific selection process in the Eneolithic communities inhabiting the Carpathians, in the context of the social differences within these communities discernible in grave furnishings. The selection was related to the age of the deceased and the visible differences in the opulence of grave goods. One of the graves in Tibava yielded twenty-five blade fragments of various lengths; this happens to be one of the richest of all the graves of the community in question, also containing gold objects, a copper adze, copper bracelets, and other artefacts.

It is reasonable to assume that in this kind of environment a method of evaluating these products will develop, which will make interpersonal communication much easier by the means of simple vocabulary. On a cognitive level, I would say, we have to do here with a magnitude and with a resource or stock that can be measured. And as the medium aspect is concerned, this is also the case of a de-linguified medium, according to Jürgen Habermas (1981), which we also have seen in the

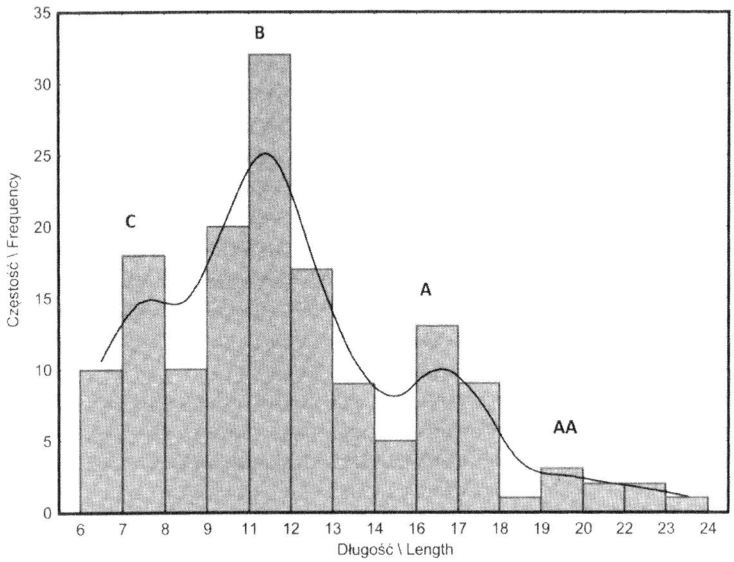

Figure 2.3. Length of flint axes made of striped flint in Poland (graphic: Dzbyński).

case of tokens. In this case, we 1) have a sought-after raw material, which is produced in a standard form in specialized workshops, and 2) we have many possibilities of further use (these tools can be processed and used several times, and shared with other people). They can therefore potentially be disposed of and stored. And they were being stored because we see in the archaeological record many hoards containing these products at this time. And finally 3) the communication aspect. The best (rationalized) way to perceive them is through simple proportions, such as a half and a quarter (the half of a half). So both media, tokens and blades, have actually the potential for discharging the language by reducing the message to number and measure (Dzbyński 2008b).

Slightly different methods of evaluation were used when it comes to bigger tools like axes. We can describe them as 'biographies of objects'. Some researchers also speak about the 'life of artefacts', where there are some stages, nexus points, which can also be points of measure (Verhart 2012). We can therefore say that every axe has a certain biography or life, which is characterized by up to four stages. A perfect example are axes made of striped flint, which were produced by specialists, by the extraction of raw material from mining shafts on the territory of today's Poland. The axes were visually very attractive tools, often polished all over their surface to resemble as much as possible the smoothness and shine of metal products. The principle of this way of proceeding is called skeuomorphism (Taylor 1999). Smoothing axes all over the surface did not increase their performance. Axes were produced in specialized workshops near the mine and then exchanged over a very large area, even 600 km away from the sources.

It turns out, as one can see it in the diagram (Fig. 2.3), that the biographies or life stages of those axes are very clearly visible: the biggest (presumably new) axes are marked as AA. Then follow A-axes, B-axes, and C-axes — this is the last stage of an axe.

We can make here a short summary. In case of stone and flint products we have these main characteristics: 1) production in specialized workshops, 2) perceived as a valuable resource, 3) remade up to four times or 4) fragmented into four segments (macrolithic blades).

The Copper Age

What we have seen so far can also be applied to the copper industries, which brings us back to the question I asked earlier in this section: How could metal be measured without the concept of weight because this procedure of measurement wasn't known yet? First of all, there are not many examples we can find in the archaeological record, but we have to bear in mind the following facts: 1) every act of measuring is relatively short and relatively rare, and 2) the use of artefacts is much longer, and it covers all previous acts of measuring. But still, there are some very illuminating examples in which case we could even speak of the Pompeii Premise, bearing in mind, however, that the concept itself is not without controversy (Schiffer 1985).

What I have in mind when speaking about the Pompeii Premise, is known in the literature as the 'copper beads' of the Cortaillod culture. They have been found in Switzerland and France. There are two well-known examples from both countries. The find from Switzerland was discovered in the 1970s and has been discussed in many ways (Sangmeister and Strahm 1974; Ottaway and Strahm 1975). Here, I am going to focus on the latest find from Colmar in France. It was uncovered in the course of a rescue excavation in 2008 (Wörle and others 2012). The context of this discovery is a grave, though a strange one. A man was placed or even thrown into a ground pit together with a group of copper beads which were placed in different parts of the pit under the skeleton. One group was placed under his chest, another was found near his pelvis, and the

Table 2.1. Weights of the beads from Colmar.

	N	Mean
Group 1	32	2.5
Group 2	16	8.2
Group 3	4	15.6
Group 4	2	31.1

Figure 2.4. All beads from Colmar could have been made from a single rod of copper (graphic: Dzbyński).

Figure 2.5. Cutting trace on one copper bead of the Cortaillod type (photo by the author, with thanks to Bernisches Historisches Museum).

last one at his feet. I measured them, as I always did, for the sake of statistical analysis, and it turns out that the deceased was equipped with four 'weight categories' which are visible in the table:

He has four beads of the second group under his chest. He also has twenty-four of the smallest pieces, and only those, near his pelvis, whereas the rest of them (all four categories) were placed at his feet. The placement of the beads is very distinctive and this could hardly be a coincidence. However, it can be somehow disturbing that we speak about the weight of these beads, since I said earlier it's not the weight we should talk about; nevertheless, this is actually the only way to measure them and to analyse them statistically. Detailed statistical investigations have been published in two papers (Dzbyński 2008b; 2014).

Thus we know how the beads were perceived by their makers. And how they were made was evidenced at the early stage of research conducted on the beads from Seeberg in Switzerland (Sangmeister and Strahm 1974). They were cut from a copper rod and twisted for carrying them on a string. I was able to observe some unfinished cuts in the museum in Bern where some of these beads are stored (Fig. 2.4). So the basic method was rather simple: you take a strip of metal and cut it proportionally according to a linear measure. In the case of Colmar, this can be done from a single rod or strip, like in Figure 2.5. It is the same method that we have seen in the case of flint macrolithic blades, but you can cut many more portions of copper than from a flint blade.

The question is: What can be measured this way? Actually a lot of things. Colmar could have been an act of measuring a copper axe. Altogether, the beads from Colmar weigh about 400 g, which happens to be the weight of an average axe from this period (Wörle and others 2012). So the axe could have been 'translated' into beads which took the form of the following portions (Table 2.1): thirty-two × quarters, sixteen × halves, four × three-quarters, and two × wholes. It is hard to say why this set of denominations was made. But every bead represented a small portion of copper, measured according to an accepted procedure, and could be worn, exchanged, and hoarded, i.e. deposited in the ground, as in the case of the beads from Seeberg which were found in a house deposit. The beads therefore were a 'measure of their own' (Budziszewski 2006).

From today's point of view, it seems strange that four typological weight categories were used in this Eneolithic community. However, the finds from Switzerland and France will no longer be strange to us if we put them in the right historical, economic, and technological context. As we have seen earlier in this chapter, in the field of flint technology related to the production of macrolithic blades, we also faced a situation where blades were fragmented appropriately. The relevant fragments could reflect specific metrological values, such as half, quarter, three-quarters, etc. It was not much different with the Colmar beads, which are like a literal transfer of these manipulations into metal. Literal has a double meaning in this case, because it also means that the perception and verbalization of these portions of raw material, both in the case of macrolithic blades and in the case of beads must have been the same or similar. The individual bead sizes were perceived not by means of an abstract weight measure but by linear, spatio-visual, and more concrete measures, because the Eneolithic communities in Europe were still on their way to a new conceptualization of metal and the full discovery of its miraculous properties. Before adequate forms of description of metal were developed, it must have been assessed similarly to stone or flint, as well as being treated in a similar manner in terms of processing (Strahm 1994).

In order to get a better picture, we can look at artefacts which are a bit easier to construct, but which show the same principle of measuring copper at an early stage. Quite recently, while browsing through the findings from the Basatanya cemetery, I came across an interesting set of fragmented wires (Fig. 2.6). In one of the rich graves of the cemetery, six wire fragments of the following sizes were discovered: 1 × 9.5 cm; 2 × 7 cm; 2 × 5 cm; and 1 × 2.5 cm. It is not difficult to read these proportions if we take the longest wire as the initial reference. So we have: 1; 2 × ¾; 2 × ½; and 1 × ¼. The

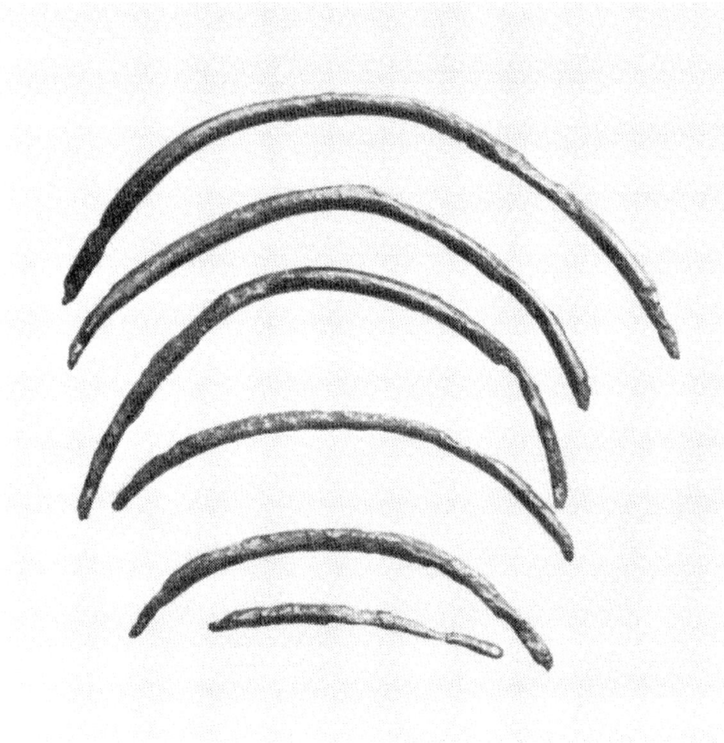

Figure 2.6. Proportionally fragmented wires of copper in the Basatanya cemetery (photo by Dzbyński).

longest wire is half the length of the standard macrolithic blades that were in circulation at that time. On the other hand, most of the blades at the Tiszapolgar sites rarely exceed 10 cm in length, like the ones deposited in the grave mentioned (Bognar-Kutzian 1963, grave 23).

We can therefore make a summary concerning early copper objects. In this case we have to do with the following phenomena: 1) specialized production, 2) valuable resource, 3) measurable as a stick or wire, 4) fragmented into four portions, but 5) larger numbers are possible, and 6) it is never-ending. By never-ending I mean the plasticity of copper. It can be used again and again, actually eternally, infinitely. This feature of metal was a revolutionary element that determined the overall perception of copper for the future. Metal did not disintegrate into useless small fragments like flint or stone. Even the smallest amount of it became valuable. This is a complete reversal of perspective. For a flint knapper, a small fragment of raw material was of no importance at all. For the metallurgist, on the contrary. Every small fragment counted, and the assessment of what could still be made of it, perhaps after adding some other small fragments, was important. 'Even slag was treasured' (Taylor 1999).

The Bronze Age

In the Bronze Age, we find evidence of 'standardization of weight' in the archaeological record, for which the earliest examples are ring bars or ring ingots (*(Ösen) ringbarren*). In the mid-1990s Majolie Lenerz De-Wilde (1995) investigated these typical findings of the Bronze Age. After analysing dozens of deposits from Austria, the Czech Republic, Germany, and Poland, it turned out that the weight of most of the items clearly oscillates around similar values, often 160–80 g (Fig. 2.7). The author also noticed that with time new types of bars appeared which are much lighter. Their weight is usually between 70 and 100 g. Smaller specimens are called *Spangenbarren* in the literature. Lenerz De-Wilde interprets this as a growing demand for bars, the creation of a market in which a kind of denomination appeared that perhaps resulted in more and more 'retail' transactions, for which smaller units of weight were needed.

At the site of Ragelstorf another interesting metrological peak was discovered (Neugebauer 2002). After analysing the deposit with statistical methods, it turned out that two peaks were emerging in the histogram: 180–90 g and 370–80 g. The first peak is very common and was observed in many cases by Lenerz De-Wilde. The second peak is rare, but in combination with the previous evidence referred to above, it clearly shows a basic metrological rule. We are dealing here with doubling or — depending on the chosen perspective — halving a portion of metal. This idea is logical (see the discussion below), commonsensical, and it was proposed by Elisabeth Pühringer for the Bronze Age (2005) and by me for the Eneolithic (Dzbyński 2004). Recently, this kind of Bronze Age portioning on casting cakes was presented by Bianka Nessel (2014). So, we are eventually getting more evidence for this kind of measuring metal in Europe. Taking into account Lenerz De-Wilde's previous arrangements for larger and smaller bars, plus the deposit from Ragelsdorf, we can imagine the whole sequence of the Early Bronze Age as follows 1–2–4 or: ½–1–2; or: ¼–½–1. We do not know how exactly the individual portions were described; the important point is, however, that this sequence is nothing else but a transformation of a linear proportion, already observed in Eneolithic materials, for example in Cortaillod beads, into portions of metal, perhaps already perceived as weight (Fig. 2.7).

As we can see, there is no third portion in this sequence, a unit that we would describe as three-quarters from today's point of view. This is an important observation, which is also visible in the case of the Corded Ware vessels, which I analysed many years ago (Dzbyński 2004). In my opinion, the invisibility of this portion is due to the transfer of manipulation from solid materials such as flint to flexible materials such as ceramics and copper. In the case of the Cortaillod beads, the presence of this portion is explained by the early age of the find and its intertwining with the cognitive structures of the Eneolithic. After all, a three-quarters long (no matter how it was called) blade was still a concrete tool for work or for exchange, whereas the ring bars were only ornaments and portions of metal. So in the case of the Corded Ware pottery and the ring bars of the Early Bronze Age, we should expect a greater abstraction in the perception of measure, which would concern respectively capacity or weight.

However, the basic question is whether the bars of the Early Bronze Age are really weight standards? Were they perceived in this way? It is worth noting that there is still no evidence in the archaeological record for the use of balance scales at that time. The first balance scales appeared in Europe in the Middle Bronze Age, several hundred years later (Pare 1999; Rahmstorf 2010; Rahmstorf this volume).

It is worth considering the hypothesis that we are, at this point, not yet dealing with weight perception systems, both in the cognitive and social sense. Some of the ring bars have a kind of seal, so-called cuffs or wire wrappings, around them. The cuffs have been interpreted as weight corrections. However, this does not seem likely. These few grams of metal added to the individual pieces were impossible to check on a precise balance scale. If these 'corrections' were some kind of addition to the transaction, they were too small, and it is hard to believe that such precision was required, especially since we do not have confirmation that such accurate metrological systems existed in Europe at that time. Yet, the statistical analyses clearly show that the weight range of these objects was very wide and *c.* 68 per cent of them weigh between 150 and 200 g (*Ösenringbarren*) or 70 and 110 g (*Spangenbarren*). As Sabine Kutscher indicates (in this volume), the cuffs were often to repair cracks in the ring bars, so they were not meant to correct the weight. That shows that the ingots were not precisely weighed, but that they were aiming to produce ingots of similar length (30 cm ± 2–4 cm). So, they were rather 'standard shapes', as Kutscher proposes. In other words, their form was still the point of reference for assessing the value of these items. Their value was still partly fixed by their form and perhaps still drew on the perception and vocabulary established in the previous period, as I suggested above in this text. So, we can say that they were 'shape-measures' of metal. But in the same way, it can be said that they were advanced forms of the sheets and wires which were in use in the Eneolithic, and actually the shape of these bars resembles quite well the wires and sheets of the previous period. When taking a look at the weight statistics of these bars, we can thus see that not only their shape but also the

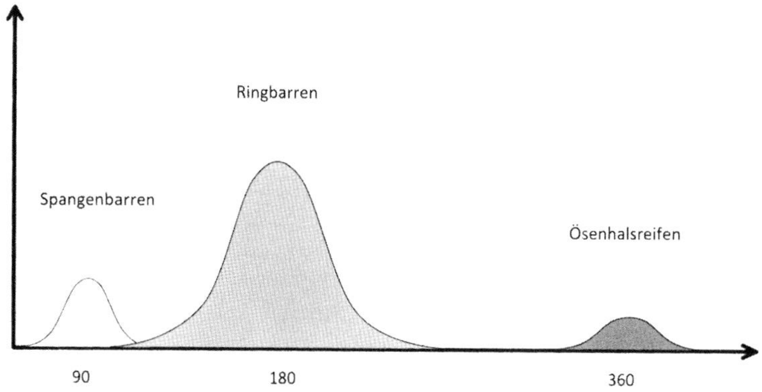

Figure 2.7. Metrological structure of the Early Bronze Age (graphic: Dzbyński).

proportions, their metrological structure, resembles the previous period.

Discussion: The Problem of Weight

The evidence so far shows that precise weighing on a scale started quite suddenly in the first half of the third millennium BC in Mesopotamia, Syria, and Egypt. We do not exactly know how the invention of the scales came about, but the findings so far suggest that it was somehow related to the emergence of state organisms in the locations mentioned.

In his book on cognitive archaeology, Colin Renfrew (2007) reflected on weight measurement as one of the evolving issues in this field, and recalls in this context the discovery of weights in the form of stone cubes from the Indus Valley. Renfrew places them in the area of symbolism, and issues the thesis that these artefacts were used to symbolize and quantify a certain 'inherent property' that had not yet been recognized. He asks: How could something like weight have happened in the first place? He states that weight had to be experienced primarily by the body. The weight could therefore be sensed by holding heavy objects and stating that they are heavier than others.

> If you have such a symbolic relationship, the stone weight has to relate to some property that exists out there in the real world. In a sense these stone cubes serving as weights are symbolic of themselves: the weight as a symbol of weight. It may be appropriate here to use the term constitutive symbol, where the symbolic or cognitive elements and the material elements co-exists. The one does not make sense without the other. (Renfrew 2007, 117)

Renfrew's statements seem to be quite general in nature, although they somehow touch the essence of the problem. It is necessary to consider, to start with, how the symbolic elements should be linked to the material elements. First of all, it is not so easy to find a stone weight in relation to something in the real world without having a proper mathematical scale. This will certainly be a symbolic element. Imagine a stone. A stone can be about as heavy, or about two or three times as heavy, as something I am just holding in my hand, and it has probably been a long time since there was no need to measure more accurately (Denny 1986). However, the exact measurement of the weight must have appeared at some point in history, probably when its mathematical symbolic element, i.e. abstract numbers, was formed. Stone cubes from the Indus Valley, like early weights from Egypt and the Middle East, suggest that it was a matter of cultural, urban-type development.

As far as Europe is concerned, the basic work was done by Christopher Pare, who published the findings of weights which, in his opinion, referred to the Aegean system. Recently, Lorenz Rahmstorf studied the phenomenon of weighing in the Bronze Age in the Old World. In his opinion, there is still a lot of work to be done in this area. By studying European and Middle Eastern weights closely, he has come to the conclusion that the range of possible shapes of the weights probably demonstrates the lack of a central political authority that could have achieved more standardization in morphology as well as in units (Rahmstorf 2010; Rahmstorf in this volume). These conclusions once again refer to the issue of a relevant political organization.

Weights and weighing in Europe are indeed emerging at a time when developed metallurgy as well as more complex Bronze Age communities appear, although these were still pre-state communities (Krause 2003, 257–62). The question that returns like a boomerang in the context outlined above is: What are the ring bars of the Early Bronze Age? As I have argued above, we should rather not interpret them as 'weight standards', because they have not yet been fully valued by measuring their weight — by a process of accessing value that began a few hundred years later in Europe. They were portions of metal, which were still perceived more visually, by their form, shape, and apparentness (*Augenscheinlichkeit*), rather than by weighing on a balance scale. The weighing scale is a technical device that seems to be simple, just as counting on fingers seems to be simple to us, and yet studying history we know that the stage of learning to count was very long. The experience of weight itself is a form of apparentness (we see that an object is heavy or light), resulting from the nature of human perception of things, but it is also a form of bodily experience difficult to quantify, just as pain, taste, or saturation of light. It is therefore an apparentness that cannot be defined without an appropriate cognitive structure to serve as a reference scale. The simplest and the earliest measuring scale is a linear, visual, and apparent scale, which has its origin in the conceptualization of anthropomorphic units such as arm, hand, or body length. How to connect an arm with a linear scale and a linear scale with weight? Logically speaking, first the simplest proportions had to be conceptualized: a whole, a half, and a quarter. As we have seen, this structure is currently evidenced in the Eneolithic.

In his text on the genesis of weight, Renfrew is speaking about 'weight as a symbol of weight', which touches the essence of the problem. Weight is an abstract measure, a measure in itself because it is technically separated from the apparentness of measure, and must appear as a certain synthesis and accumulation of values. Together with the above findings, this should guide us toward asking the right questions: How did the line or the linearity idealization take place? How was the linear scale created and how does it relate to the sensual experience of the weight of specific objects? As to where to search further for answers to these questions, important guidance came from Endmund Husserl who wrote several important works on the sources of mathematics. First of all, his works *Über den Begriff der Zahl* (1887) and *Philosophie der Arithmetik: Psychologische und Logische Untersuchungen* (1891) should be mentioned here. However, he gave the most general and accessible picture of the philosophy and history of mathematics in his final work *The Crisis of European Sciences* (1936). In this book he makes a detailed analysis of the historical formation of the process of mathematization as geometrization in the first place, which is a derivative of the practical need to measure the world. Using his own method of phenomenological reduction, Husserl believes that in the world that surrounds us we first experience various 'non-ideal' bodies — i.e. geometrically non-ideal. Geometry and the sciences connected with it have to do with space-time bodies and shapes, figures, also movement shapes, deformation changes, etc. which are possible in space-time, especially as measurable quantities. It becomes now clear that even if we don't know much about the historical surroundings of the world of the first geometricians, so much is certain as an unchangeable, essential structure: it was a world of things and of artefacts. It was still a pre-geometric world, and this pre-geometric world was a world of things disposed of according to an 'an exact space and time' (Derrida 1989). In his essay, Husserl writes that since the geometry inherited by Galileo was already *techne*, far from the sources of real, apparent thinking, and since this was the geometry for the ancient Greeks, we have to look back to the origins of ancient history in order to search for the sources of this apparentness. We have to find these senses of the first apparent sources of geometric thinking in action and thinking of our prehistoric ancestors. In the conceptualization of weight we can see this transformation of geometric and mathematical thinking, about which Husserl writes in his essay. The weight seems to be the pivotal problem, where on the one hand the experiences ('real praxis') of the real life of prehistoric societies converge, and at the same time it is the beginning, the starting point for the 'ideal praxis', in which only ideal shapes and ideal numbers are being further transformed. If the previous experiences 'converge', as I put it, it means that there must have been an earlier stage. Taking Husserl into account, we should assume that this earlier stage was the phase of the 'real praxis'.

Practical application of the balance scale for measuring weight requires the idealization of the measuring rod out of the original contexts of apparent linear measure, and the use of a purely mathematical aspect of a geometric line. On the one hand, we have here the ideal rod halves, which form the weight arms and the geometrically constructed linear scale. The invention of the balance scale is covered with a veil of prehistory and there is little evidence on this subject. However, given the above-mentioned theory, we can say that the balance scale must bear, on the one hand, a connection with the idealization of linearity and the idealization of the numerical linear scale, and on the other hand, with the phenomenological, sensual experience of weight as a measure. In other words, weighing on a balance scale is the application of a certain theory concerning the measuring rod. It is the application of the measuring rod in a different aspect than its original, tangible, anthropomorphic linear measurement. The practice of using the rod as a linear measure, as it logically follows, was primordial in relation to the theory that was developed on the basis of these earlier experiences. The invention of the balance scale must, logically speaking, result from the previous experience of measuring in an anthropomorphic way, because the balance scale gets rid of the bodily aspect, isolating the person from the object of the measuring experience. It is the technical device that takes the measurement, and the person only looks at it and 'understands' the operation of the device.

Imagine this from the perspective of metal technology, because it seems that metallurgy could have played an important role in the invention of the balance scale. Weight makes it possible to idealize the measure of metal and more precisely calculate the proportion of the alloy. This means that, eventually, two metals must be treated as ideal quantities and added to each other in theory before they are added in practice. Of course, in the early stage there had to be metallurgical experiments concerning what proportions should be used in a given alloy. The first alloy in the history of the world was bronze (a copper-tin alloy), and in the archaeological material we do indeed find evidence that such experiments took place (Strahm 1994; Krause 2003). So we get something that we can describe as real praxis, and the consequences: ideal praxis. Let us add to this a thought experiment: in the real praxis situation, i.e. before the invention of the weight measurement, we have different objects of different shapes, and we cannot embrace them with one measure. A linear measure does not allow for direct reference to each other for wire, sheet metal, axe fragments, or other tools. In such a procedure, even if the manufacturer has the appropriate experience, there will always be a lack of precision and predictability. It is only by applying a common measure to all objects, and using a common scale — weight — that practical idealization, cognitive blend of different shapes and different objects becomes possible. This is why, as the archaeological material suggests, it is in the Bronze Age that we can speak of the beginning of the 'ideal praxis', ultimately leading to pure mathematics. Thanks to this, a metallurgist can calculate exact proportions of each new object, calculate the demand for the future, construct ideal models in thought, that is, basically, make some kind of thought experiment in the world of ideal objects. Thanks to this, new conceptual approaches are possible, like the idea of an unlimited number of things that can be made by the procedure (see Seaford in this volume). However, before this happened, we shall not forget about the world of 'real praxis', the world in which calculations were not so simple, and some were even impossible.

Works Cited

Bellingwood, P. 2005. *First Farmers: The Origins of Agricultural Societies* (Malden: Wiley-Blackwell).

Bognar-Kutzian, I. 1963. *The Copper Age Cemetery of Tiszapolgar-Basatanya*, Archaeologica hungarica, 42 (Budapest: Akademiai Kiado).

Budja, M. 1998. 'Clay Tokens. Accounting before Writing in Eurasia', *Documenta praehistorica*, 25: 219–35.

———. 2003. 'Seals, Contracts and Tokens in the Balkans Early Neolithic: Where in the Puzzle', *Documenta praehistorica*, 30: 115–30.

Budziszewski, J. 2006. 'Flint Economy in Chalcolithic Societies of East-Central Europe', *Veröffentlichungen aus dem Deutschen Bergbau-Museum Bochum*, 148: 315–27.

Chapman, J. 2000. *Fragmentation in Archaeology: People, Places and Broken Objects in the Prehistory of South Eastern Europe* (London: Routledge).

Chapman, J., T. Higham, V. Slavchev, B. Gaydarska, and N. Honch. 2006. 'The Social Context of the Emergence, Development and Abandonment of the Varna Cemetery, Bulgaria', *European Journal of Archaeology*, 9.2–3: 159–83 <https://doi.org/10.1177/1461957107086121>.

Damerow P., R. K. Englund, and H. J. Nissen. 1988. 'Die Ersten Zahldarstellungen und die Entwicklung des Zahlbegriffs', in B. Riese (ed.), *Schrift und Sprache* (Heidelberg: Spektrum Akademischer Verlag), pp. 102–11.

Denny, P. J. 1986. 'Cultural Ecology of Mathematics. Ojibway and Inuit Hunters', in M. P. Closs (ed.), *Native American Mathematics* (Austin: University of Texas), pp. 129–80.

Derrida, J. 1989. *Edmund Husserl's Origin of Geometry: An Introduction* (Lincoln, NE: University of Nebraska Press).

Dunbar, R. 1992. 'Neocortex Size as a Constraint on Group Size in Primates', *Journal of Human Evolution*, 22.6: 469–93 <https://doi.org/10.1016/0047-2484(92)90081-J>.

———. 1998. *Grooming, Gossip, and the Evolution of Language* (Cambridge, MA: Harvard University Press).

Dzbyński, A. 2004. *Metrologische Strukturen in der Kultur mit Schnurkeramik und ihre Bedeutung für die Kulturentwicklung des mitteleuropäischen Raumes*, Beiträge zur Ur- und Frühgeschichte Mitteleuropas, 39 (Halle: Beier & Beran).

———. 2008a. *Ritual and Understanding: Rational Bases of Communication and Exchange in Prehistoric Central Europe* (Rzeszów: Mitel).

———. 2008b. 'Von Seeberg bis Kelsterbach: Ein Beitrag zur Bedeutung des Kupfers im Äneolithikum und in der Bronzezeit Europas', *Prähistorische Zeitschrift*, 83: 36–44 <https://doi.org/10.1515/PZ.2008.002>.

———. 2011. 'Mr Blademan. Macrolithic Technology – Eneolithic Vocabulary and Metaphors', *Documenta praehistorica*, 38: 172–84 <https://doi.org/10.4312/dp.38.14>.

———. 2014. 'From Seeberg to Colmar: Early Mathematical Concepts in Prehistoric Europe at the Interface between Material Culture, Technology and Metaphors', *Praehistorische Zeitschrift*, 89.1: 1–11 <https://doi.org/10.1515/pz-2014-0001>.

Emmons, G. T. 1991. *The Tlingit Indians* (Seattle: University of Washington Press).

Graeber, D. 2011. *Debt: The First 5000 Years* (Brooklyn: Melville House).

Habermas, J. 1981. *Theorie des kommunikativen Handelns*, I: *Handlungsrationalität und gesellschaftliche Rationalisierung* (Berlin: Suhrkamp).

Hayden, B. 1995. 'A New Overview of Domestication', in T. D. Price and A. Gebauer (eds), *Last Hunters – First Farmers: New Perspectives on the Prehistoric Transition to Agriculture*, (Santa Fe: School of American Research Press), pp. 273–99.

Husserl, E. 2017. *Kryzys nauk europejskich i fenomenologia transcendentalna* [The Crisis of European Sciences and Transcendental Phenomenology] (Kraków: Vis-à-vis Etiuda).

Jopling, C. F. 1989. 'The Coppers of the Northwest Coast Indians: Their Origin, Development, and Possible Antecedents', *Transactions of the American Philosophical Society*, 79.1: 1–164.

Krause, R. 2003. *Studien zur kupfer- und frühbronzezeitlichen Metallurgie zwischen Karpatenbecken und Ostsee* (Rahden: Leidorf).

Kristiansen, K. 1991. 'Chiefdoms, States, and Systems of Social Evolution', in T. Earle (ed.), *Chiefdoms: Power, Economy and Ideology* (Cambridge: Cambridge University Press), pp. 16–43.

Kristiansen, K., and T. Larsson. 2005. *The Rise of the Bronze Age Society* (Cambridge: Cambridge University Press).

Krukowski, S. 1922. 'Pierwociny krzemieniarskie górnictwa, transportu i handlu w holocenie Polski' [The Beginnings of Flint Mining, Transportation and Trade in Holocene Poland], *Wiadomości Archeologiczne*, 7: 1–34.

Kula, W. 2004. *Miary i ludzie* (Warsaw: Książka i Wiedza).

Lakoff, G., and R. Núñez. 2001. *Where Mathematics Comes from: How the Embodied Mind Brings Mathematics into Being* (New York: Basic Books).

Lenerz De-Wilde, M. 1995. 'Prämonäre Zahlungsmittel in der Kupfer- und Bronzezeit Mitteleuropas', *Fundberichte aus Baden-Württemberg*, 20: 229–327.

Lichardus, J. 1991. 'Das Gräberfeld von Varna im Rahmen des Totenrituals des Kodžadermen-Gumelnita-Karanovo VI-Komplexes', in J. Lichardus and R. Echt (eds), *Die Kupferzeit als historische Epoche: Symposium Saarbrücken und Otzenhausen 6.-13.11.1988*, I (Bonn: Habelt), pp. 167–94.

Lichter, C. 2004. *Untersuchungen zu den Bestattungen des südeuropaischen Neolithikums und Chalkolithikums* (Darmstadt: Von Zabern).

Macnaughtan, D. 2020. 'Native American Coppers of the Northwest Coast', *Academia Letters* <https://doi.org/10.20935/AL108>.

Malafouris, L. 2010. 'Grasping the Concept of Number: How Did the Sapient Mind Move beyond Approximation?', in C. Renfrew and I. Morley (eds), *Archaeology of Measurement: Comprehending Heaven, Earth and Time in Ancient Societies* (Cambridge: Cambridge University Press), pp. 35–42.

———. 2013. *How Things Shape the Mind: A Theory of Material Engagement* (Cambridge, MA: MIT Press).

Manolakakis, L. 2005. *Les industries lithiques énéolithiques de Bulgarie* (Rahden: Leidorf).

McLuhan, M. 1964. *Understanding Media: The Extensions of Man* (New York: McGraw Hill).

Michailidou, A. 2010. 'Measuring by Weight in the Late Bronze Age Aegean: The People behind the Measuring Tools', in C. Renfrew and I. Morley (eds), *The Archaeology of Measurement Comprehending Heaven, Earth and Time in Ancient Societies* (Cambridge: Cambridge University Press), pp. 71–87.

Nessel, B. 2014. 'Bronze Age Portioning of Raw Metal – Concepts, Patterns and Meaning of Casting Cakes', *Apulum: Acta Musei Apulensis*, 51: 401–25.

Neugebauer, J.-W. 2002. 'Die Metalldepots der Unterwölblinger Kulturgruppe Ragelsdorf 2 und Unterradlberg 1 und 2. Überlegungen zum prämonetären Charakter der niedergelegten Wertgegenstände', *Mitteilungen der Anthropologischen Gesellschaft in Wien*, 132: 25–40.

Ottaway, B., and C. Strahm. 1975. 'Swiss Neolithic Copper Beads: Currency, Ornament or Prestige Items?', *World Archaeology*, 6.3: 307–21 <https://doi.org/10.1080/00438243.1975.9979611>.

Overmann, K. 2018. 'Constructing a Concept of Number', *Journal of Numerical Cognition*, 4.2: 464–93 <https://doi.org/10.5964/jnc.v4i2.161>.

Pare, C. 1999. 'Weights and Weighting in Bronze Age Central Europe', in *Eliten der Bronzezeit: Ergebnisse zweier Kolloquien in Mainz und Athen* (Mainz: Verlag der Römisch-Germanischen Zentralmuseums).

Pühringer, E. 2005. 'Als es weder Kilos noch das Einmaleins gab: Gedanken über ein urzeitliches Masssystem als Ordnungsprinzip', *Experimentelle Archäologie in Europa*, 3: *Bilanz 2004*: 15–23.

Rahmstorf, L. 2010. 'The Concept of Weighing during the Bronze Age in the Aegean, the Near East and Europe', in I. Morley and C. Renfrew (eds), *The Archaeology of Measurement: Comprehending Heaven, Earth and Time in Ancient Societies* (Cambridge: Cambridge University Press), pp. 87–105.

Renfrew, C. 1990. *Archaeology and Language: The Puzzle of Indo-European Origins* (Cambridge: Cambridge University Press).

———. 2007. *Prehistory: The Making of the Human Mind* (London: Phoenix).

Sangmeister, E., and C. Strahm. 1974. *Die Funde aus Kupfer in Seeberg, Burgäschisee-Süd*, Acta Bernensia, 2 (Bern: Stämpfli).

Schiffer, M. B. 1985. 'Is There a "Pompeii Premise" in Archaeology?', *Journal of Anthropological Research*, 41.1: 18–41 <https://doi.org/10.1086/jar.41.1.3630269>.

Schmandt-Besserat, D. 1992. *Before Writing: From Counting to Cuneiform* (Austin: University of Texas Press).

———. 2007. *Jak powstało pismo* (Warsaw: AGADE).

———. 2010. 'The Token System of the Ancient Near East: Its Role in Counting, Writing, the Economy and Cognition', in I. Morley and C. Renfrew (eds), *The Archaeology of Measurement: Comprehending Heaven, Earth and Time in Ancient Societies* (Cambridge: Cambridge University Press), pp. 27–34.

Sériadès, M. L. 2000. 'Spondylus Gaederopus: Some Observations on the Earliest European Long Distance Exchange System', in S. Hiller and V. Nikolov (eds), *Karanovo*, III: *Beiträge zum Neolithikum in Südosteuropa* (Vienna: Phoibos), pp. 423–37.

Šiška, S. 1964. 'Pohrebisko tiszapolgarskiej kultury w Tibave', *Slovenska archeologia*, 12: 293–356.

Strahm, C. 1994. 'Anfänge der Metallurgie in Mitteleuropa', *Helvetia archaeologica*, 25: 2–39.

Taylor, T. 1999. 'Envaluing Metal: Theorizing the Eneolithic "Hiatus"', in S. S. M. Yong, A. M. Pollard, P. Budd, and R. A. Ixer (eds), *Metals in Antiquity*, British Archaeological Reports International Series, 79 (Oxford: Archaeopress), pp. 222–32.

Tomasello, M. 2002. *Kulturowe źródła ludzkiego poznania* [The Cultural Origins of Human Cognition] (Warsaw: Państwowy Instytut Wydawniczy).

Verhart, L. 2012. 'Neolithic Stone Tools and the Transition from Mesolithic to Neolithic in Belgium and the Netherlands, 5300–4000 cal BC', *Journal of Archaeology in the Low Countries*, 4.1: 5–35 <http://jalc.nl/cgi/t/text/get-pdf4734.pdf?c=jalc;idno=0401a01> [accessed 1 August 2021].

Wörle M., S. Van Willigen, C. Strahm, M. B. Merkl, E. Hildbrand, F. Chenal, R.-M. Arbogast, and P. Lefranc. 2012. 'Inhumations, dépôts d'animaux et perles en cuivre du IVe millénaire sur le site Néolithique récent de Colmar "Aérodrome" (Haut-Rhin)', *Bulletin de la Société préhistorique française*, 109.4: 689–730 <https://doi.org/10.3406/bspf.2012.14203>.

FRANZISKA FECHER

3. Feasting and Burial Rites in Pre-Hispanic Honduras

Comparing Archaeological and Ethnographic Evidence

ABSTRACT Excavations at the coastal settlement of Guadalupe (AD 900–1525), north-east Honduras, uncovered evidence of feasts held in association with burials. An integral part of these feasts was the consumption of food and drink and the food remains were deposited in the vicinity of the burials. These festivities were carried out repeatedly, resulting in the accumulation of a dense concentration of ceramics, bones, and other objects over time. The findings further indicate the possibility that burials were reopened and bones were removed for secondary use. Such practices of feasting and secondary use of burials are mentioned in ethnographic sources. The text discusses the extent to which parallels can be drawn between the archaeological record and written sources and addresses feasting as an important social practice in pre-Hispanic Honduras.

KEYWORDS north-east Honduras, Cocal period, ethnographic sources, feasting, social interaction

Introduction

Honduras is located on the Central American land bridge. The narrow land bridge links the large land masses of North America and South America and represents an important corridor between the two continents. In pre-Hispanic times, the present-day territory of Honduras occupied a special position on this connecting bridge. For it was here that three major culture areas intersected: Mesoamerica, southern Central America, and the Caribbean (Fig. 3.1). Although culture areas are constructs of archaeological research and should be understood as highly simplistic concepts, it is nevertheless apparent that different cultures, traditions, and ways of life met and came into contact with each other in this specific region. Despite this particular circumstance, little archaeological research has been conducted in Honduras. Research has focused primarily on the study of the Maya city of Copan in the western, Mesoamerican part of the country. Copan's exploration dates back to the middle of the nineteenth century. At that time, the drawings of the sunken ruins in the dense Central American jungle by architect and painter Frederick Catherwood fascinated a large audience in the USA and Europe (Stephens [1841] 1969).

Franziska Fecher (franziska.fecher@uzh.ch) is research associate at the Prehistoric Archaeology Division at the University of Zurich. She studied cultural anthropology of the Americas, prehistoric archaeology, and history of art at the Universities of Bonn, Puebla, and Zurich. Her dissertation is about networks of interaction in the Late Pre-Hispanic period (ad 900–1525) in Honduras. She took part in archaeological fieldwork in Germany, Mexico, and Honduras. Her main research interests include network analysis, ceramic analysis, interdisciplinary and comparative archaeology.

Number and Measure in Prehistory: Gift, Exchange, and the Transfer of Goods, ed. by Philippe Della Casa and Aleksander Dzbyński, WSA, 6 (Turnhout, 2024), pp. 41–56.
BREPOLS PUBLISHERS DOI 10.1484/M.WSA-EB.5.137128

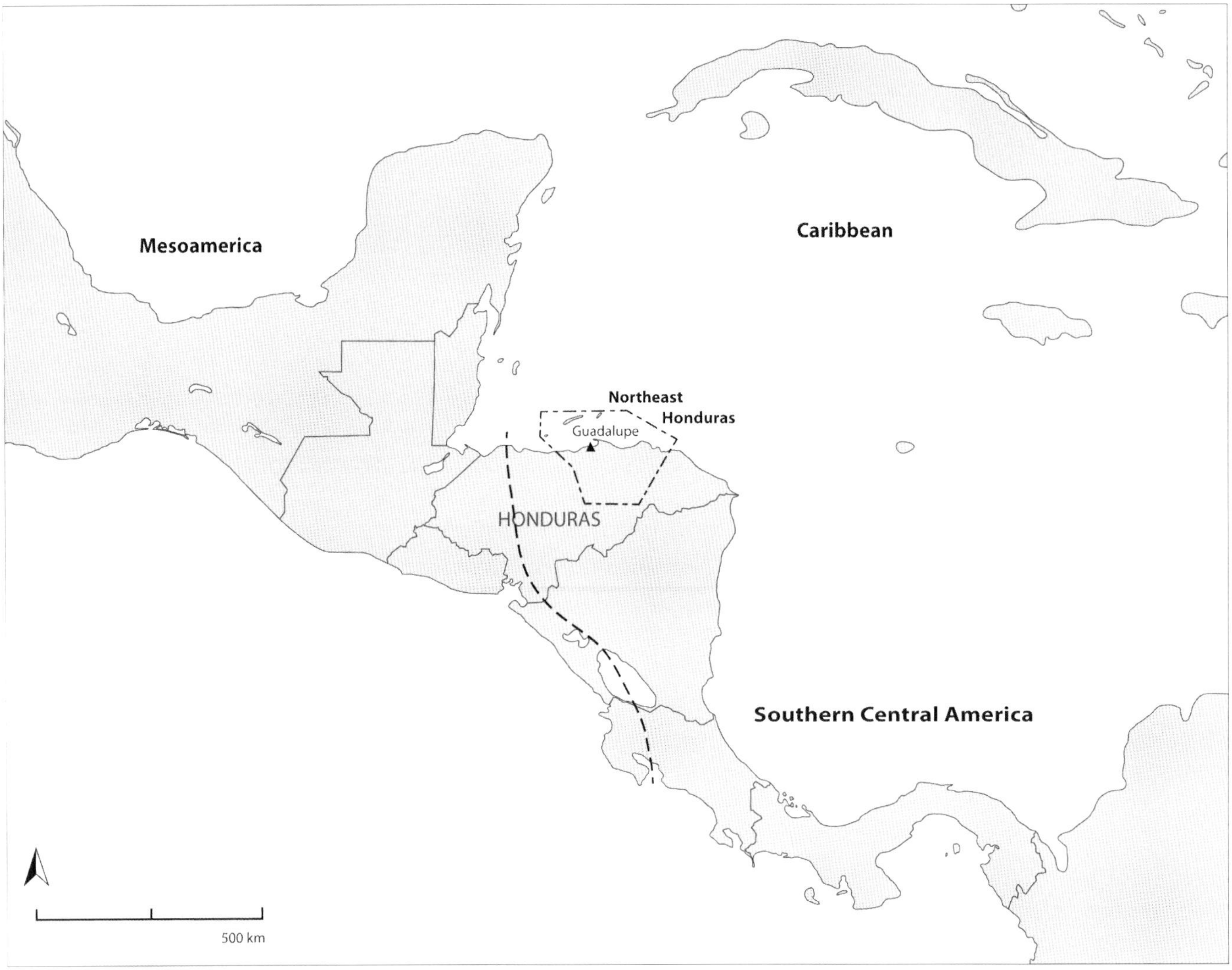

Figure 3.1. Culture areas of Middle America. The dashed line indicates the border between Mesoamerica and southern Central America (F. Fecher).

In the meantime, Copan has become one of the most intensively studied Mayan cities at all. Several projects have also been carried out in the surrounding regions in the west of the country. Prominent sites have been excavated and entire valleys have been systematically explored (e.g. Urban 1986). This contrasts sharply with the state of investigation in the centre and east of Honduras. Archaeological research has been sporadic there. Much of it was conducted by pioneers who came to Honduras from North America in the first half of the twentieth century (e.g. Spinden 1925; Stone 1941; Strong 1935). Modern research began in the 1970s with the work of Paul Healy and the first use of ^{14}C dating (Healy 1984). Important subsequent research projects include activities of the Honduran Antiquities Authority IHAH (e.g. Epstein and Véliz 1977; Hasemann 1977), research in the Culmi Valley (Begley 1999), and excavations at Selin Farm (Goodwin 2019). Although these projects have contributed important insights to understanding the pre-Hispanic history of north-east Honduras, many fundamental questions still remain open. For example, there is a seven-hundred-year gap in the chronology (between 400 BC and AD 300), for which initial evidence was only found during excavations in 2020 (Reindel and others 2021). Typologies of find material are little elaborated or hardly available, especially for the early periods. Little information exists on the economic system, on the structure and function of settlements, or on burial rites. Furthermore, for a long time the image of central and eastern Honduras as a periphery dominated, in contrast to the 'complex societies' that had developed in Mesoamerica. At the same time, cultural achievements were primarily attributed to external influences by some researchers (Begley 1999; Lara Pinto and Hasemann 1988).

Since 2016, archaeological research has been undertaken to clarify central questions of pre-Hispanic cultural history and to investigate its internal processes. As a result, the performance of feasts associated with burials was identified at the coastal settlement of Guadalupe (AD 900–1525). The more detailed study of the processes involved in carrying out the festivities represents an important step toward a better understanding of the pre-Hispanic society. However, the archaeological remains present limitations in terms of possible interpretations and inferences. In the following text, the definition of feasts is addressed and possibilities of their investigation in the archaeological record are considered. Subsequently, the archaeological finds and features from north-east Honduras are compared with ethnohistorical sources in order to reconstruct associated activities and reasons behind the festivities. It is shown that feasts were an important practice in pre-Hispanic north-east Honduras and still play an important role in indigenous societies today.

Feasting in Pre-Hispanic Societies

The topic of feasts has received more attention in archaeological theory only in the last decades. A significant step in this respect was certainly the emergence of the post-processual approach, which was interested in individual processes within societies and attributed agency to material culture itself (Hodder 1986). The 2001 edited volume by Michael Dietler and Brian Hayden helped to establish feasts as a central object of study in archaeology and to develop theoretical and methodological tools for their systematic investigation. Feasts are now recognized as a specific form of ritual activity. They are viewed as a basic form of human behaviour, the study of which enables the understanding of social and cultural processes and dynamics. Moreover, they are considered to drive the emergence of social hierarchies and early states (Bray 2003a; Chacon and Mendoza 2017).

Dietler and Hayden (2001, 3) define feasts basically as 'events essentially constituted by the communal consumption of food and/or drink'. Additionally, they point out, 'Most authors are also explicit in differentiating such food-consumption events from both everyday domestic meals and from the simple exchange of food without communal consumption' (Dietler and Hayden 2001, 3). These are the core definitions of festivities. In addition, there are numerous other elements that can accompany the communal consumption of food. Feasts are often associated with music or dances that contribute to the special and festive atmosphere. Performing ritual acts, formal speeches, dramaturgical effects, or wearing special clothing or masks can also be part of feasts. In these activities and symbols, the world-view of a society is reflected and expressed. All these components contribute to the fact that feasts are distinctive situations that differ from everyday life. Feasts provide a framework for emotionality and excitement, conventions can be disrupted (Leimgruber 2010). The use of alcohol or hallucinogenic substances contributes to such a special experience. Celebrating together, sharing emotions, and experiencing the unusual, the unconventional, strengthens the sense of community. Strengthening social ties can be a central reason for engaging in festivities (Hayden 2001, 30). At the same time, it can emphasize and reinforce the differentiation of various groups within a society.

Furthermore, political alliances can be forged during feasts, or people can become part of a social group through initiation rites. Especially for societies without writing, feasts have the important function that the agreements and understandings made during the festivities can be witnessed by those present. They thus assume the function of eyewitnesses and can confirm and also pass on what they have experienced — an important function that is taken over in our society today by documents and contracts. Feasts often frame important social or political transactions and can have a strong demonstrative and representative character. The display of wealth and power of the organizer can be in the foreground and the participation in excessive festivities can lead to the participants being indebted to the host. Thus, feasts also represent a special form of social exchange that is usually based on reciprocity.

The occasions for holding feasts are numerous. Junker (2001, 271) shows that feasts are often associated with important points of political-economic life (warfare, events associated with the agricultural cycle, chiefly succession, etc.) or with stages of the human life cycle, such as birth, marriage, illness, and death. Thus, they also serve as important instruments to remember past events. If they are performed periodically, they also divide time into periods and help to structure everyday life.

Numerous ethnohistorical sources attest to the fact that feasts played a central role in many indigenous societies. The best-known example is probably the *potlatch*, which is celebrated by indigenous groups on the north-west coast of Canada and the USA. These celebrations centre on the passing on and affirmation of name titles, the (re)affirmation of social and political connections, and the simultaneous display of power and prestige. This practice is particularly impressive and revealing, as there are numerous descriptions and studies of the festivities that provide detailed insight into the meaning and purposes as well as the organization and procedure of the festivities (e.g. Jonaitis 1991; Piddocke 1965). The fact that the *potlatch* is a living tradition, still celebrated today, enables a unique access to the deeper meaning of this social event.

There is also ample evidence of the importance of feasts in indigenous societies in Central America. For the Maya culture, whose southernmost extensions are found in Honduras, the importance of feasts is known from the archaeological record, including scenes on polychrome vessels, and from colonial period sources. Diego de Landa (Tozzer 1941, 92) describes two different types of feasts. Feasts were celebrated among the elite, where cocoa acted as the main ceremonial drink, precious gifts were exchanged, and, in a kind of reciprocal system, the guests of the feast were indebted to host a similar feast in the future. The second type of feasts is described among the commoners and usually took place on the occasion of important life cycle events. In all feasts, the consumption of alcoholic beverages (e.g. the fermented maize drink *chicha*) apparently played an important role.

For southern Central America, María Eugenia Bozzoli (1979) gives an interesting insight. Ethnohistorical sources and her own observations among the Bribri, an ethnic group living in Costa Rica, testify to excessive festivities that can last several days or even weeks. Because they involve the consumption of large quantities of alcoholic beverages, they are also known as *chichadas*. They are often associated with funerals, and some sources report that the food of the deceased is consumed during the festivities. Also here, Bozzoli (1979, 144) interprets these festivals as feasts of reciprocity. For the same indigenous group, William Gabb (1875, 500–03) describes the celebrations of the burial of King Santiago. These celebrations extended over two to three weeks and were accompanied by excessive banquets, dances, and songs. In addition, he observes that several *chicha* containers were broken on purpose during a procession, at the same time he observed that 'nothing of real value was destroyed' (Gabb 1875, 502). These accounts illustrate that feasts occupied and still occupy an important place for many indigenous societies. Due to the changes and Christian influence that have occurred since the colonial period, we certainly cannot directly apply these observations to pre-Hispanic societies. However, the archaeological record can also provide hints on the performance of feasts. Hayden (2001, table 2.1) summarizes in a table characteristics that can be detected in the archaeological context and indicate the holding of festivities. These include the presence of unusual food and pottery vessel forms, such as an unusually high quantity of them, the presence of feasting facilities, prestige items, and ritual items, as well as the identification of these features at special locations.

In recent years, a growing number of archaeologists have interpreted features as feasting debris, which previously were mainly referred to as middens or refuse heaps (McOmish 1996). This new interpretive approach provides access to a variety of information that can be retrieved from the study of the archaeological record. For example, it can be reconstructed which social groups or households were involved in the preparation or execution of festivities (e.g. Hendon 2003; Vaughn 2004). Information about the architecture and infrastructure that was associated with the holding of festivities can also provide information on the subject (e.g. Cook and Glowacki 2003; Knight 2001). Often, studies of festivities also involve a functional examination of the pottery, providing information about which vessel forms were used for which purposes (e.g. Bray 2003b; Goodwin 2019; Smith 2003; Vaughn 2004). It can be found out what kind of foodstuffs were eaten during celebrations through botanical or zooarchaeological studies (e.g. Kelly 2001; Knight 2001). On a more theoretical level, these observations in turn allow us to draw conclusions about the structure of society, political strategies, organization of power, and forms of exchange (e.g. Goldstein 2003).

However, there are also aspects of the festivities that can hardly be reconstructed from the archaeological remains, such as the sequence of activities during the festivities, invited guests, or even motivations and beliefs related to the celebrations. In this case, consulting written sources is a way to better understand the archaeological remains and gain insight into the practices and background of the festivities that studying archaeological remains alone cannot reveal. However, the interpretation of the archaeological record based on ethnographic sources has its limits and must be done with caution. It certainly cannot be assumed that an unaffected continuity exists between the social practices leading to archaeological remains and observations on indigenous groups described in ethnographic sources. The arrival of Europeans in the Americas caused a significant disruption and change of social behaviour and especially religious traditions have been influenced by Christian beliefs. Nevertheless, ethnographic data can, if critically applied, offer access to different interpretation models. Thus, in the following text different ethnographic sources on indigenous societies in north-east Honduras will be compared to the archaeological record and used to reconstruct a possible scenario that led to the archaeological remains identified in Guadalupe and in other sites of north-east Honduras.

The Archaeological Record from North-East Honduras

Investigations at Guadalupe

Since 2016, archaeological research has been taking place on the Caribbean coast of Honduras in collaboration with Swiss, German, and Honduran institutions to investigate the pre-Hispanic history of this previously

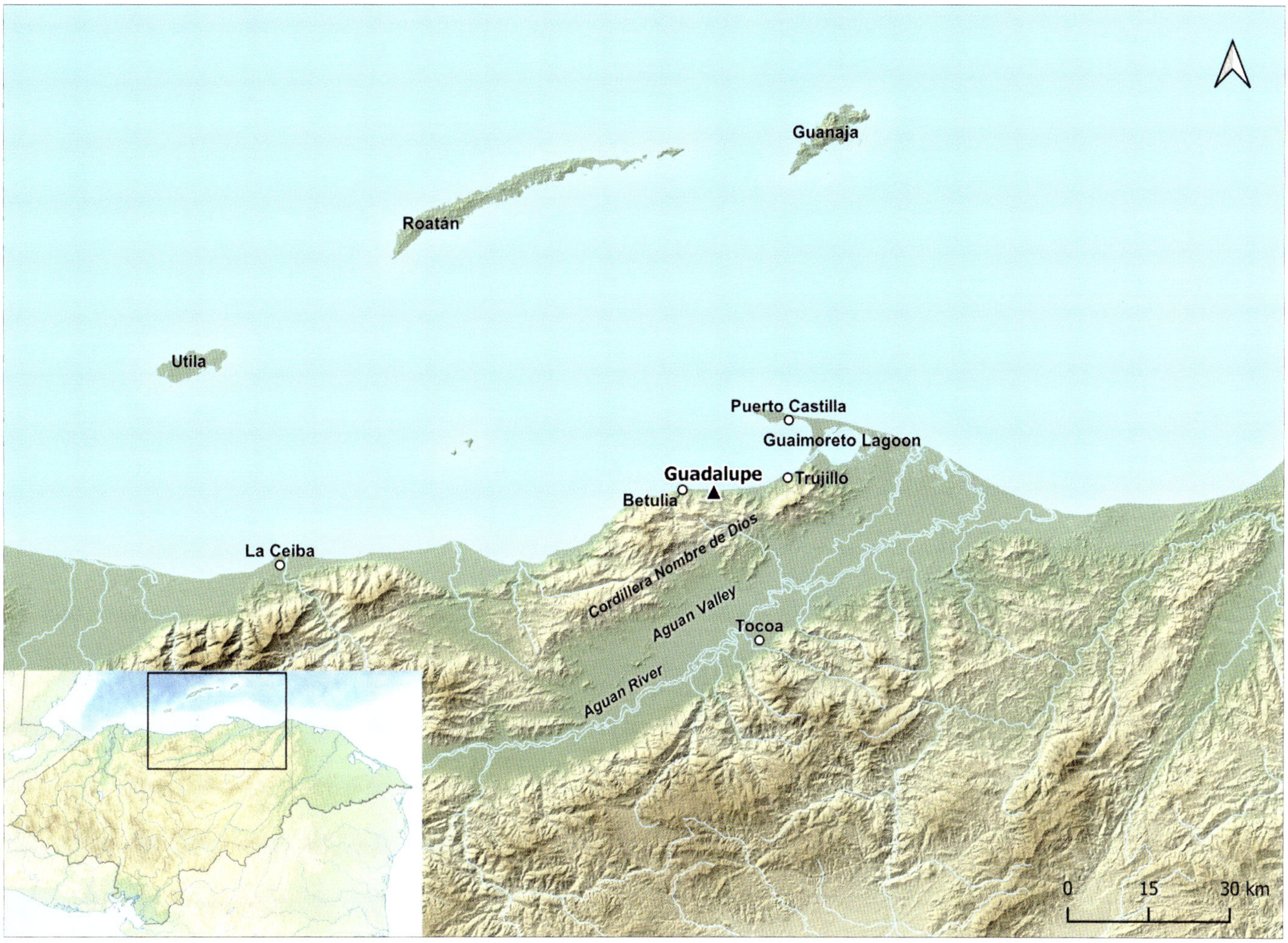

Figure 3.2. Location of Guadalupe on the Atlantic coast of Honduras (M. Lyons, F. Fecher).

little-known region and to shed new light on its relationship with surrounding regions (Fecher 2023; Fecher and others 2020; Reindel and others 2020). For the Late Pre-Hispanic period (AD 900–1525) in particular, research from neighbouring Mesoamerica shows that exchange along the coast played an important role (Sabloff and Rathje 1975). In order to clarify the role of this coastal region in interregional exchange relations, investigations have focused in particular on the coastal region between Trujillo to the east and Betulia to the west, as well as the adjacent Aguan Valley to the south (Fig. 3.2). This area contains numerous archaeological sites that are visible above ground but have not yet been studied. The following descriptions refer in particular to the coastal site of Guadalupe, which for several reasons occupies a special position within the investigated settlement structure.

Guadalupe is located directly on the Atlantic coast, about 15 km west of Trujillo (Fig. 3.3). The pre-Hispanic remains are located on a terrace that extends towards the coast as an extension of the coastal cordillera (Fig. 3.4). Today, the archaeological remains are largely overbuilt by modern construction, but concentrations of ceramic sherds are found throughout the terrace surface and are evidence of pre-Hispanic settlement activities. Only a few architectural elements are superficially preserved, because in the region buildings were mainly constructed with perishable material. Today, traditional houses are built with cobble stones that serve as basis for a structure of wattle and daub (*bajareque*). Roofs are made of palm leaves. Findings from Guadalupe suggest that this construction method was also used in pre-Hispanic times (Mattes 2019). Superficially, such structures are hardly identifiable. However, a clearly recognizable earthen mound is located in the centre of the mentioned terrace (Fig. 3.5). Today this area belongs to the terrain of the local primary school. The mound measures 20 m in diameter and 1 m in height and has an approximately rectangular shape. In the south-west it is cut by the construction of a modern road. To clarify the composition and stratigraphy of the mound, a 2 × 12 m trench was dug into the mound in a

Figure 3.3. Aerial photo of the modern village of Guadalupe (M. Reindel).

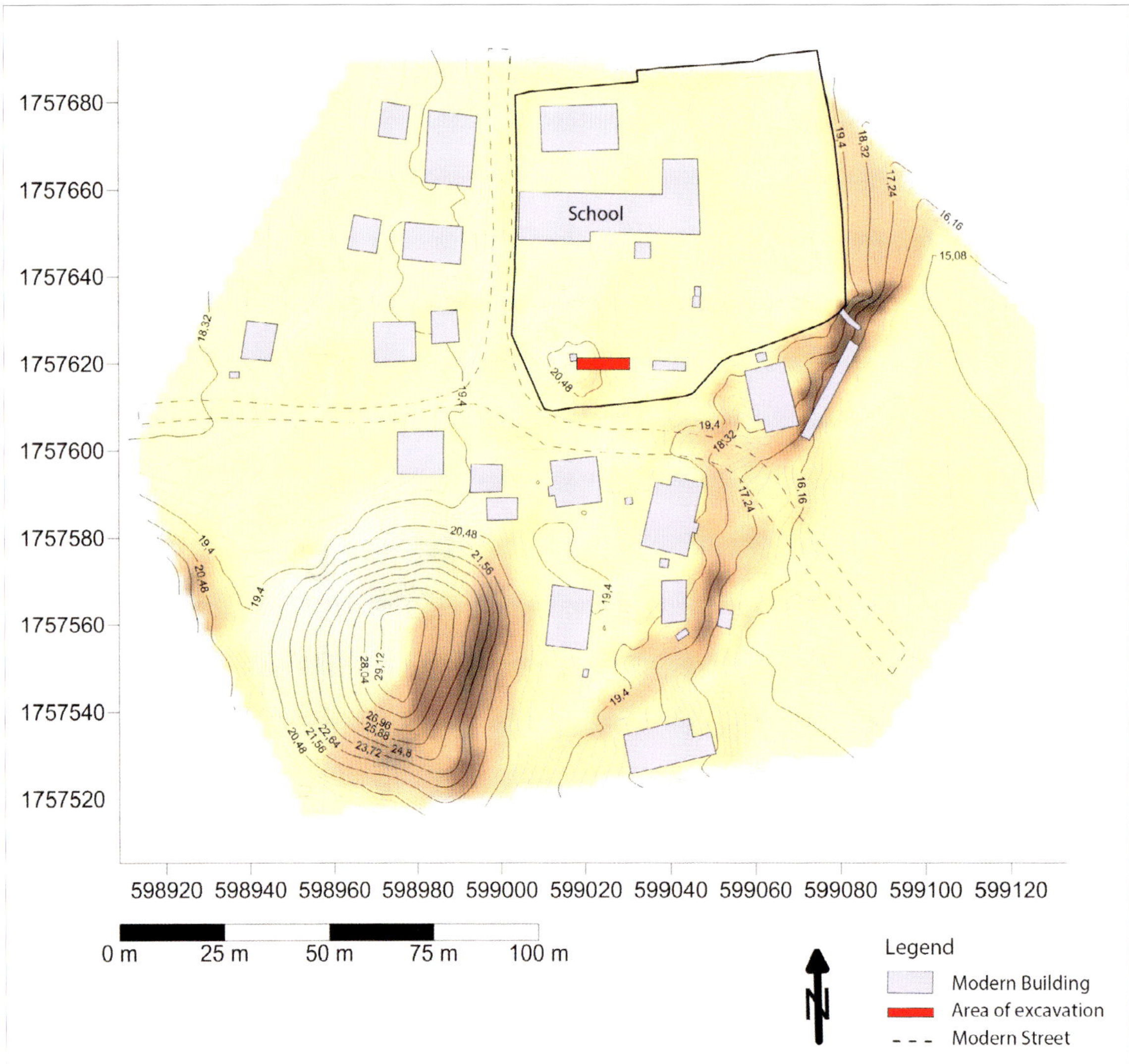

Figure 3.4. Topographic map of Guadalupe (M. Lyons, M. Reindel, F. Fecher).

west–east direction, with the western section marking the centre of the mound and the eastern end extending to the base of the mound. The results of the excavation indicate a period of use from AD 900 to 1525 at the latest and show two clearly distinguishable phases (Fig. 3.6).

The upper layers are composed of densely packed objects, which are about 1 m thick in the centre and thin out towards the sides (Fig. 3.7). The find material consists to a large extent of pottery fragments. It is remarkable that these do not present small fragments, as one would expect within a settlement, but they are pottery broken into large pieces. In many cases a large percentage of the vessels could be reconstructed (Fig. 3.8). In addition, some vessels were apparently broken *in situ*. Associated with the pottery were other finds such as stone implements, shells, animal bones, obsidian tools, metal artefacts, and greenstone objects, all surrounded by dark brown, humic earth. No clear internal stratigraphy could be detected within the accumulation, although colour differences of the soil indicate different layers. Due to the particular composition of the find assemblage, this feature is interpreted as the remains of ritual activities, which involved the consumption of food and the subsequent deposition of the vessels and food remains, together with other objects.

The reason for this debris being located in this specific location became clear when the excavation process continued. Underneath the find concentration, several accumulations of *bajareque* were identified. Below these concentrations of *bajareque*, burial pits with interments were identified. All burials are located in the periphery of the mound, where the find concentration thins out (Fig. 3.9). The skeletons of the burials were in a rather poor state of preservation and none of the graves contained grave goods. The skeletons of at least five individuals were identified. Two individuals are located in the centre of the plana with east–west orientation and were completely excavated (Fig. 3.10). Individual 1 was found in an extended dorsal position. Physical anthropological studies under the direction of Julio Chi Keb and Luisa Mainou (Chi Keb 2020; Mainou and Aguilar Romero 2020) showed, that the anatomical position of the bones indicates that the deceased was buried in a kind of box or coffin or was wrapped in a cloth, leading to the shoulders being slightly elevated, and the arms lying close to the body. The position of the articulated bones further indicates a primary burial. However, not all bones were present. Notably, the pelvis, femurs, and right arm were missing. The cranium was also no longer present; only individual fragments were found near the sternum, which may indicate that the head was once supported but fell forward when the organic material perished.

The second individual was located to the east of Individual 1. It was found in a semiflexed lateral position. Also in this case, several bones were missing, such as the pelvis, the thorax region, and parts of the arms. A third individual was located in a burial pit extending into the north profile, which is why only a part of the skeleton was visible in the excavation trench. The bones were in a poor state of preservation and no further studies were conducted. In addition to these three skeletons, individual human bones were found in various locations, the connection between which remains unclear. These include a skull in the south profile, a cluster of long bones, and a skull base that was level and in line with Individuals 1 and 2 (see Fig. 3.10). Below these burials, two other dark, elongated discolourations were identified that most likely also represent burial pits but have not been opened to date.

The excavations showed only a small section of the overall situation. However, given the high quantity of burials in this small excavated area, it can be assumed that there exist numerous other burials in the periphery of the mound. The find situation certainly represents a place where burials were carried out over a long period of time, which were associated with ritual activities. The debris of these activities was deposited directly next to the burials. In terms of the burials, the missing skeletal parts and the presence of individual bones are particularly striking. On the one hand, this situation could be due to specific preservation conditions, which may well affect individual bones differently (Mainou and Aguilar Romero 2020). However, in combination with the presence of single bones such as cranial parts and long bones, which were not associated with any other skeletal remains, another possible explanation comes to mind. This pattern suggests a secondary use of primary burials in which individual bones were removed. Another important observation is that the burials can be stratigraphically associated with the find concentration. This is also indicated by the analysis of ^{14}C data. This fact leads us to the interpretation that the find concentration represents the remains of ritual activities carried out in connection with the burials and their possible secondary use. The composition of the find concentration indicates the consumption of food and drink, as well as the intentional deposition of vessels that apparently served as containers for the food. Musical instruments and prestige items are also found among the finds (Fig. 3.11). Such categories of finds are referred to by Brian Hayden (2001, table 2.1) as indicative of the identification of feasts in the archaeological record. The ritual activities can thus be more accurately described as feasts, and the find concentration can be interpreted as feasting debris. This results in the picture of a site that was used over a long period of time for the burial of deceased. In

Figure 3.5. Pre-Hispanic mound in Guadalupe (M. Reindel).

Figure 3.6. Excavation in 2019. Distinguishable layers are clearly visible in the profiles (M. Reindel).

Figure 3.7. Ceramic concentration (M. Reindel).

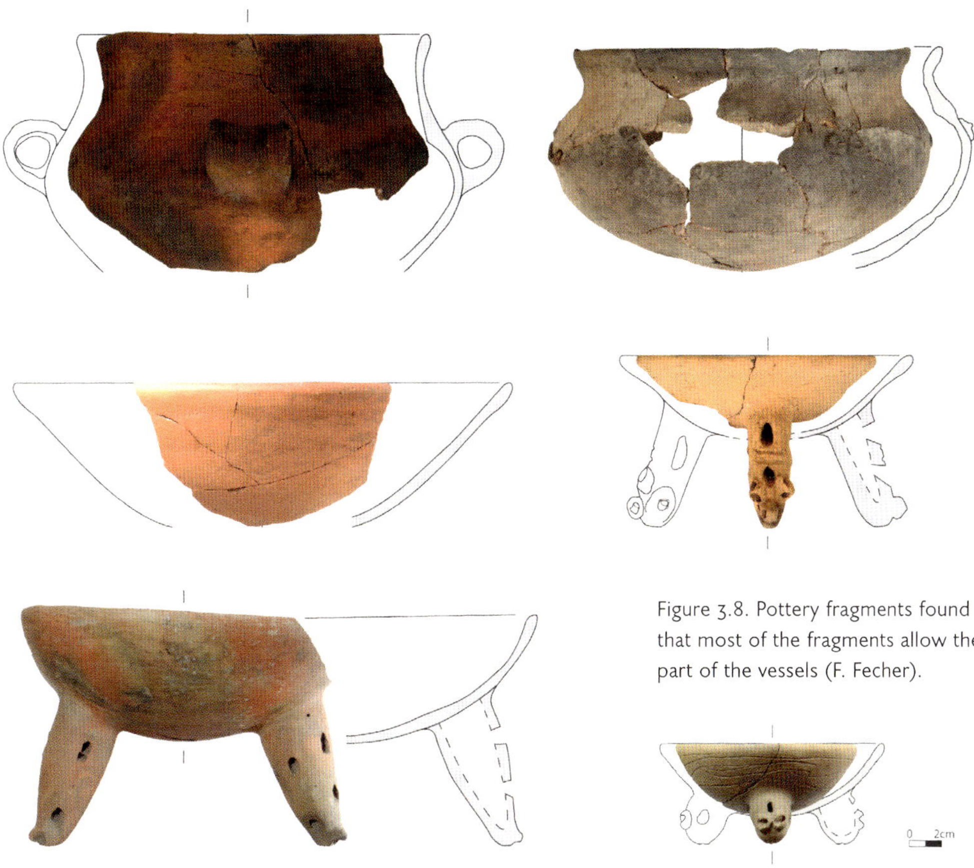

Figure 3.8. Pottery fragments found during excavations. Note that most of the fragments allow the reconstruction of a large part of the vessels (F. Fecher).

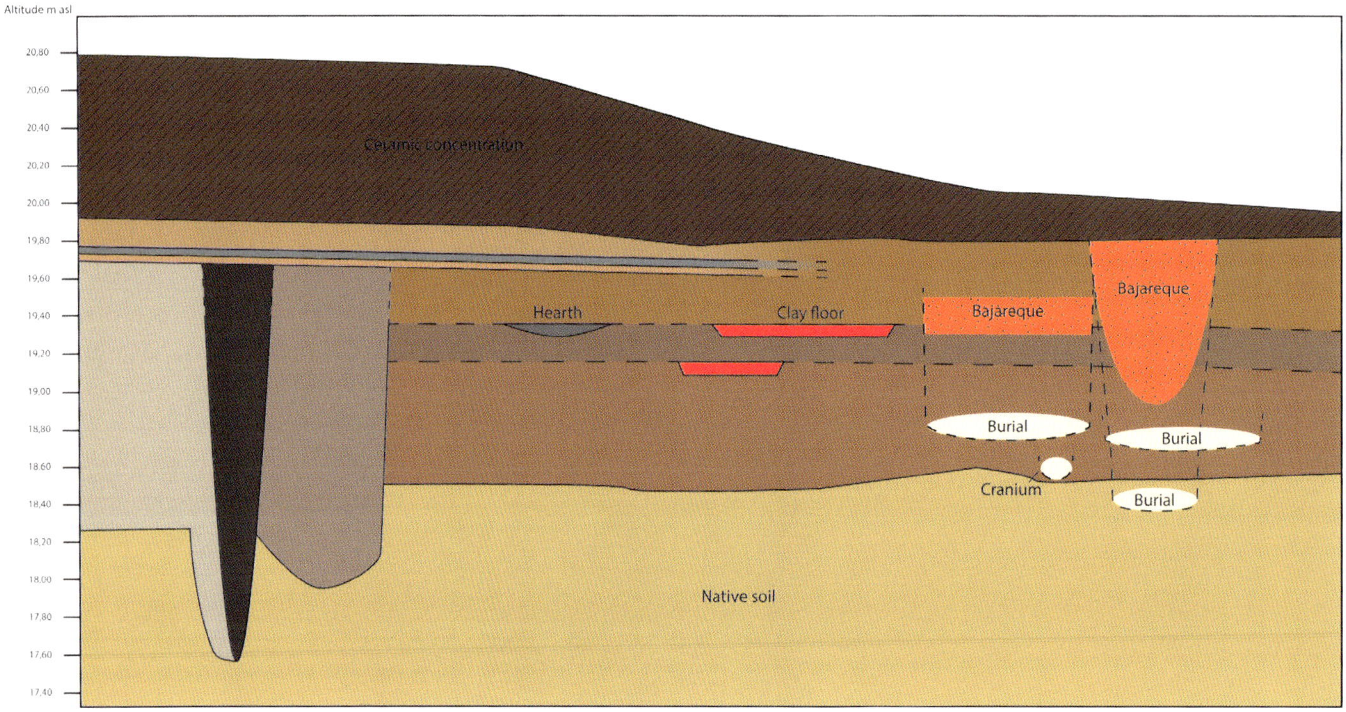

Figure 3.9. Schematic representation of the stratigraphy showing the most important features. Dashed lines indicate uncertain boundaries (F. Fecher).

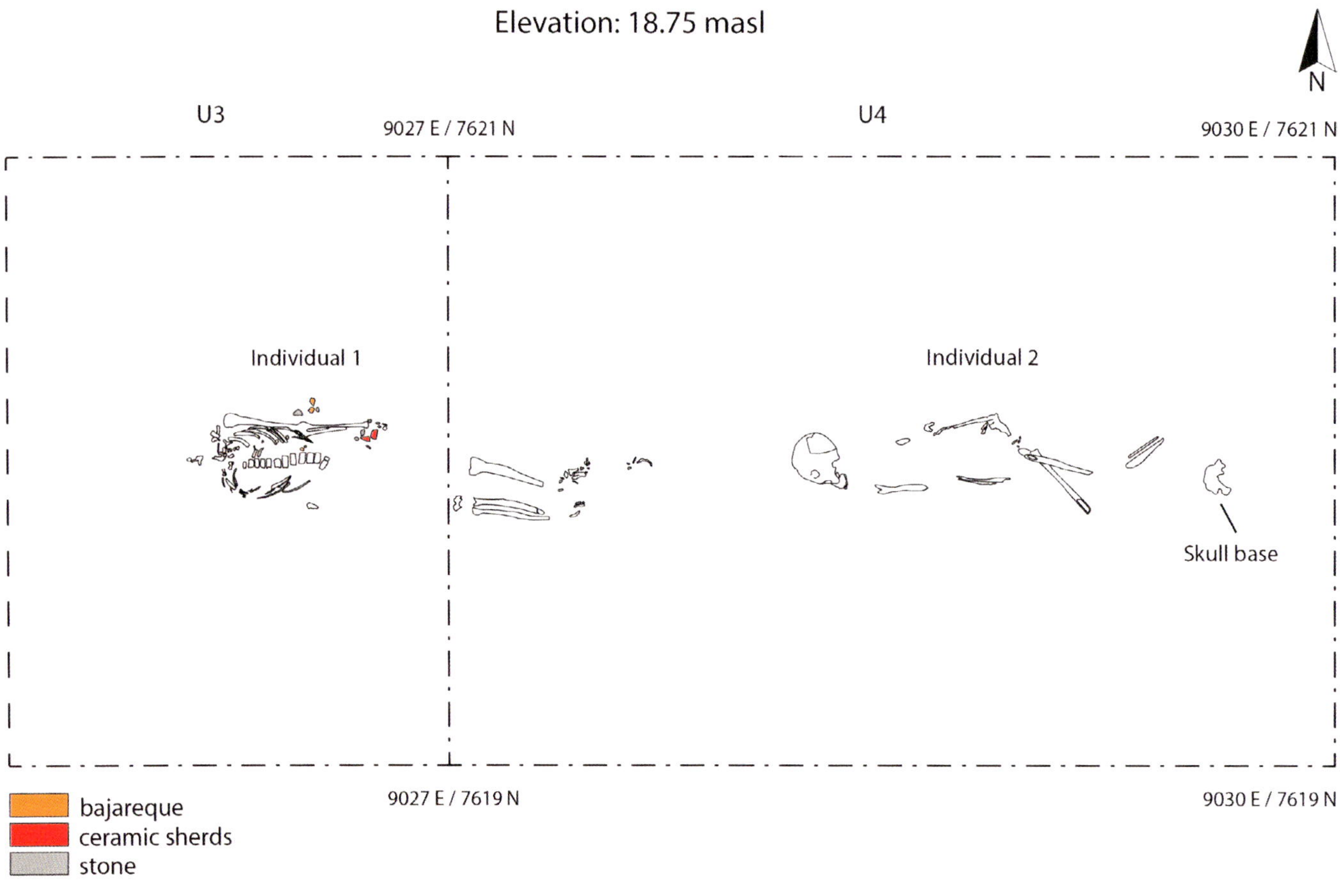

Figure 3.10. Location of burials in Unit 3 and Unit 4 at an elevation of 18.75 masl (F. Fecher).

Figure 3.11. Zoomorphic ocarinas (P. Bayer, F. Fecher).

connection with the burials, festivities were held and the remains of those were deposited on site, right next to the burials. This accumulation occupied a central place within the settlement and was visible to all its inhabitants, thus serving as a kind of marker or memorial to the deceased. The identification of feasts in Guadalupe is significant in that they represent an important social practice and can be considered as a specific form of social exchange.

Reports from Other Sites

It seems that Guadalupe was not the only site where ritual feasting has been practised in pre-Hispanic times. We find evidence of similar practices in other archaeological sites in north-east Honduras that have not previously been interpreted in this direction, but can be re-evaluated with the new findings from Guadalupe and the theoretical framework at hand. A careful study of the early excavation reports from the first half of the twentieth century reveals striking parallels. We have to keep in mind that the excavation standards were different during this time and that information on features is not always as detailed as would be desirable. Nevertheless, interesting observations were recorded, on which the findings from Guadalupe shed new light.

The description for Indian Hill Site 1 on Barbereta Island by Strong (1935, 86–111), is very reminiscent of the situation found at Guadalupe. Indian Hill Site 1 consists of an approximately 1 m thick accumulation of closely packed ceramics measuring 20 m in diameter. Strong emphasizes that the ceramic fragments were broken into large pieces, allowing for the reconstruction of some of the vessels. The pottery was mixed with shells, animal bone, and stone tools. Although no burials were found, Strong discovered individual human teeth and bones, indicating the presence of human burials, which might not have been discovered due to unsystematic excavation work. A similar situation is reported for Dixon Site on the adjacent island of Roatán. Here, Strong (1935, 51–72) describes a concentration of ceramic sherds distributed over an area of about 10 m. In the centre of the accumulation, he found an offertory cache in the form of a complete Bay Island Polychrome vessel filled with 487 objects, including greenstone beads, metal bells and other copper objects, shell artefacts, and stone celts.

Strong interprets Indian Hill 1 and Dixon Site as offertory sites. Furthermore, he assumes that the pottery accumulations were caused by the long-term and periodic use by the islands' inhabitants and that they accumulated over a long period (Strong 1935, 146). These sites show a similar situation to that identified in Guadalupe, thus they probably can be identified as feasting sites as well. Based on the pottery, the last two sites date to the Cocal phase (AD 1000–1525). That feasts in north-east Honduras also occupied an important role during the preceding Selin phase (AD 300–1000) and thus appear to have had a long tradition was recently shown by excavations in Selin Farm, where enormous deposits of broken pottery and molluscs indicate the presence of regular feasting activities (Goodwin 2019).

There are also some reports that indicate that the secondary use of skeletal remains might have been a common practice in pre-Hispanic north-east Honduras. Near Trujillo, Herbert Spinden (1925) describes the finding of eight burials located inside an earthen mound. One of these burials 'had been dismembered and the arms and legs were laid across the body'. For the Black Rock Basin site on Utila Island, Strong (1935, 22–27) describes the discovery of eight urn burials. Accompanying these urns were individual bundles of long bones, as well as individual skulls arranged around the urns. Both findings indicate the practice of a secondary use of individual bones. However, more detailed information, such as the dating of the sites, is limited. The findings from Guadalupe and the reports from other archaeological sites in north-east Honduras suggest that feasting in association with burials was an important activity during pre-Hispanic times, executed at different places starting at least in the Selin period. In the next section, the ethnographic records are consulted to provide additional information on the practice of ritual feasting.

The Ethnographic Record from North-East Honduras

A look at the ethnographic documents from north-east Honduras reveals significant parallels with the archaeological evidence from Guadalupe and the other sites. The examination of the written sources helps us to better understand the reasons for the festivities and the burials, and to get an idea of which kind of activities might have been associated with them.

An important source are the writings of Eduard Conzemius (1927; 1928; 1932). Among others, he conducted fieldwork among the Pesh. This ethnic group inhabited large parts of north-east Honduras when the Spanish arrived, and it is likely that they also inhabited the area around Guadalupe. Their population was severely decimated by the European conquest. Today, the Pesh live in small areas in the departments of Olancho, Gracias a Dios, and Colón (Griffin and others 2009). Conzemius combines his own observations with information from historical reports and provides information on burial rites (1927, 300s). He relates that when a person dies, it is usual to wrap the corpse in a cloth made of tree bark. A canoe serves as a coffin in which the person is buried. The coffin is then placed in the house where the deceased had lived and the house is subsequently abandoned. If several people die within a short time, it occurs that the entire settlement is abandoned. As grave goods, the deceased obtain objects that they used during their lifetime, such as tools, weapons, or tobacco. These objects are to accompany them on their journey to paradise. However, only 'items that are broken or of little use'[1] are buried with the dead (Conzemius 1927, 300). In the past, it was also usual to kill the deceased's animals and to destroy his fruit trees and cultivations.[2] The funerals are accompanied by excessive festivities. Conzemius (1927, 301) describes these gatherings as 'orgies that are accompanied by strange dances, food and chicha'.[3] In addition to celebrations on the day of the death, festivities take place after three, nine, and thirty days. Finally, after one year, the main feast is held.

Further important information comes from the writings of Anne Chapman (1958, 132–39). She makes similar interesting observations by evaluating various historical accounts especially from the seventeenth to nineteenth centuries. Like Conzemius, she mentions that the dead were often buried in buildings that are subsequently abandoned. In some cases, buildings are even built over the graves. She also mentions the use of boats as coffins. She also summarizes reports of Pesh festivities that testify to the fact that such events were accompanied by the constant drinking of alcoholic beverages and numerous rituals.

Recent ethnographic studies on the Pesh prove the importance of funeral celebrations (Griffin and others 2009, 78–81). They describe that during funerals, ceremonies are performed by a shaman. These include the preparation and consumption of specific foods and drinks such as *chicha* made from corn or manioc and the provision of food for the spirits of the deceased. To this end, various instruments are played to call upon the spirits. On the ninth day after the day of death, a feast is held where the village community gathers to consume food and drink and dance together. The remains of the festivities are then collected and deposited at a specific location.

Regarding the Sumu and the Miskito, who also live in north-east Honduras today, John Esquemeling (1924, 237) describes a secondary use of human remains: if a man dies, he is buried. After a year, his widow opens the grave to remove a portion of the bones. The bones are cleaned and dried. Afterwards, the woman carries the bones with her in a bag for one year. Although caution is required when linking these observations to the findings in Guadalupe, as we currently assume that the Sumu did not inhabit the area around Trujillo in pre-Hispanic times and that the Miskito are an ethnic

1 Translation from Spanish by the author. Original text: 'artículos que están quebrados o de poca utilidad'.

2 For the Miskito, where he documents a similar procedure, he explains this behaviour as follows: 'This is in order to prevent the deceased from being robbed by the living, thus preventing his or her spirit to linger about doing all sorts of mischief' (Conzemius 1932, 155).

3 Translation from Spanish by the author. Original text: 'orgías que son acompañadas de bailes extraños, comida y chicha'.

group that only formed after the Conquista, this report still offers an interesting perspective on the findings in Guadalupe.

Discussion

The ethnographic sources provide information about the process of the feasting activities and the ideas and beliefs behind them. Some of the elements described above certainly go back to Christian-influenced ideas, such as preparing the deceased for their journey to paradise or using a rosary for prayers (Conzemius 1927, 300; Griffin and others 2009, 79). At the same time, we find elements in the ethnographic sources that can also be identified in the archaeological record in Guadalupe and which might date back to pre-Hispanic times. First of all, one parallel is that funeral celebrations play an important role in both contexts and that they were held in certain intervals. The burials and the assemblage of feasting debris and intentionally deposited objects occupy a central position in the pre-Hispanic settlement of Guadalupe. The accumulation of objects must have represented a prominent point within the settlement and must have been visible to the inhabitants at all times. The ^{14}C data show that the accumulation occurred over several years, indicating that festivities were conducted periodically.

Interesting, too, is the connection of burials to architecture, which is described in the sources. The concentration of *bajareque* found at Guadalupe associated with the burials indicates that also in the pre-Hispanic context, there was a certain connection between burials and architecture although it is difficult to reconstruct if the burials were located beneath houses or if the *bajareque* concentrations are the result of the piling up of *bajareque* remains above them. The use of a coffin or cloth to wrap the corpse also finds parallels in the archaeological record, for the analysis of Individual 1 indicated such a practice. Recent studies mention that food leftovers from the feasts were collected and deposited at a specific location. Interestingly, the deceased was given broken objects or things of little value, and his possessions were destroyed. Guadalupe's deposit may reflect this partial destruction of objects and the deposition of both the deceased's possessions and leftover food from festivities. This might explain the somewhat strange mix of food remains, tools, and valuable items such as metal and jade objects as well as the fact that no individual grave goods were found within the burial pits. The accumulation might be interpreted as a kind of common grave good for the deceased.

In addition to information on the course of events and the use of material culture, the sources also provide indications of underlying reasons for the performance of the feasts. Due to the prominent placement of the feasting debris in the middle of the settlement of Guadalupe, the dead were assigned a central place in society, both spatially and socially. The world of the living and the dead were closely related, no strict boundary was created, as is the case in today's Western world-view. Rather, the situation in Guadalupe presents a special kind of interaction and connection between the living and the dead. This closeness of the living and the deceased is known for many premodern societies (Steadman and others 1996). In these cases, the connectedness between the world of ancestors and the living is explained by the belief that ancestors still represent active agents and can certainly influence daily life (van Schaik and Michel 2017, 113). The social ties do not end with death. The belief that the spirits of the deceased are still present and are part of the community as active actors also resonates in the recent ethnographic account of the Pesh when they provide food and drink to the spirit of the deceased (Griffin and others 2009, 80) or in the account of the Miskito, who seek to keep the spirits of the dead from malicious activities through their behaviour at the festivities (Conzemius 1932, 155).

Thus, the performance of the festivities can eventually be understood as an action to appease the still powerful deceased. The feasts can be seen as a kind of gift. In return, the ancestors are expected to behave favourably toward the living. This idea in turn is based on the fact that human beings are strongly influenced by reciprocal ideas (Adloff and Mau 2005). Collaboration and mutual support are essential to human survival. In this case, human interaction and reciprocity even go beyond the world of the living and extend into the realm of the dead. At the same time, the basic reciprocal behaviour of humans, the expectation of a return for a given favour, can be understood as a driving force for the emergence of more formalized forms of exchange.

Summary

This article presented new archaeological research results from north-east Honduras, which are interpreted as remains of feasts held in relation to burials. The study of ethnographic sources has revealed significant parallels to the archaeological record and has provided new perspectives for the interpretation of the archaeological remains. In addition, it has been shown that the performance of festivities is linked to the notion of reciprocity, which in this case extends into the realm of the dead. The performance of feasts in north-east Honduras represents an important social practice, and brings us a step closer to understanding the internal processes of pre-Hispanic cultural development in this region.

Acknowledgements

The research activities on the north coast of Honduras are a cooperation between the University of Zurich, the German Archaeological Institute, the Instituto Hondureño de Antropología e Historia, and the Museum Rietberg Zurich. They are led by Markus Reindel, and receive funding from the Swiss-Liechtenstein Foundation for Archaeological Research Abroad (SLSA). To all these institutions and persons I express my sincere thanks. In addition, I would like to thank the UZH Alumni's Research Talent Development Fund (FAN) for the research grant, and Mike Lyons, Markus Reindel, and Paul Bayer for providing the illustrations. I would also like to thank Luisa Mainou and Julio Chi Keb for the physical anthropological analyses, and all the participants of the Guadalupe/Colón projects.

Works Cited

Adloff, F., and S. Mau (eds). 2005. *Theorie und Gesellschaft*, LV: *Vom Geben und Nehmen: Zur Soziologie der Reziprozität* (Frankfurt: Campus).

Begley, C. 1999. 'Elite Power Strategies and External Connections in Ancient Eastern Honduras' (unpublished doctoral dissertation, University of Chicago).

Bozzoli, M. E. 1979. *El nacimiento y la muerte entre los Bribris* (San José: Editorial Universidad de Costa Rica).

Bray, T. L. (ed.). 2003a. *The Archaeology and Politics of Food and Feasting in Early States and Empires* (New York: Kluwer).

———. 2003b. 'To Dine Splendidly: Imperial Pottery, Commensal Politics, and the Inca State', in T. L. Bray (ed.), *The Archaeology and Politics of Food and Feasting in Early States and Empires* (New York: Kluwer), pp. 93–142.

Chacon, R. J., and R. G. Mendoza (eds). 2017. *Studies in Human Ecology and Adaptation*, VIII: *Feast, Famine or Fighting? Multiple Pathways to Social Complexity* (Cham: Springer).

Chapman, A. 1958. 'An Historical Analysis of the Tropical Forest Tribes on the Southern Border of Mesoamerica' (unpublished doctoral thesis, Columbia University).

Chi Keb, J. 2020. 'Informe Proyecto Guadalupe Colón, Honduras' (unpublished report).

Conzemius, E. 1927. 'Los Indios Payas de Honduras: Estudio geográfico, histórico, etnográfico y lingüístico', *Journal de la société des Americanistes*, 19: 245–302.

———. 1928. 'On the Aborigines of the Bay Islands (Honduras)', in *Proceedings of the Twenty-Second International Congress of Americanists*, II (Rome), pp. 57–68.

———. 1932. 'Ethnographical Survey of the Miskito and Sumu Indians of Honduras and Nicaragua', *Bureau of American Ethnology Bulletin*, 106: 1–191.

Cook, A. G., and M. Glowacki. 2003. 'Pots, Politics, and Power: Huari Ceramic Assemblages and Imperial Administration', in T. L. Bray (ed.), *The Archaeology and Politics of Food and Feasting in Early States and Empires* (New York: Kluwer), pp. 173–202.

Dietler, M., and B. Hayden (eds). 2001. *Feasts: Archaeological and Ethnographic Perspectives on Food, Politics, and Power*, Smithsonian Series in Archaeological Inquiry (Washington, DC: Smithsonian Institution Press).

Epstein, J. F., and V. Véliz. 1977. 'Reconocimiento arqueológico en la Isla de Roatán, Honduras', *Yaxkin*, 2.1: 28–39.

Esquemeling, J. 1924. *The Buccaneers of America* (London: Routledge).

Fecher, F. 2023. *Archaeological Investigations in Guadalupe, Northeast Honduras: Interaction Networks during the Late Pre-Hispanic Period (AD 900-1525)*, Forschungen zur Archäologie Aussereuropäischer Kulturen, 19 (Wiesbaden: Harrassowitz).

Fecher, F., M. Reindel, P. Fux, B. Gubler, H. Mara, P. Bayer, and M. Lyons. 2020. 'The Ceramic Finds from Guadalupe, Honduras: Optimizing Archaeological Documentation with a Combination of Digital and Analog Techniques', *Journal of Global Archaeology*, 2020: 1–54 <https://doi.org/10.34780/joga.v2020i0.1009> [advance online publication].

Gabb, W. M. 1875. 'On the Indian Tribes and Languages of Costa Rica', *Proceedings of the American Philosophical Society*, 14: 483–602.

Goldstein, P. S. 2003. 'From Stew-Eaters to Maize-Drinkers: The Chicha Economy and the Tiwanaku Expansion', in T. L. Bray (ed.), *The Archaeology and Politics of Food and Feasting in Early States and Empires* (New York: Kluwer), pp. 143–72.

Goodwin, W. A. 2019. 'Communities of Consumption on the Southeastern Mesoamerican Border: Style, Feasting, and Identity Negotiation in Prehispanic Northeastern Honduras' (unpublished doctoral dissertation, Southern Methodist University).

Griffin, W., H. Escobar Martínez, and J. C. Hernández Torres. 2009. *Los Pech de Honduras: una etnia que vive* (Tegucigalpa: Instituto Hondureño de Antropología e Historia).

Hasemann, G. 1977. 'Reconocimiento arqueológico de Utila', *Yaxkin*, 2.1: 40–76.

Hayden, B. 2001. 'Fabulous Feasts: A Prolegomenon to the Importance of Feasting', in M. Dietler and B. Hayden (eds), *Feasts: Archaeological and Ethnographic Perspectives on Food, Politics, and Power*, Smithsonian Series in Archaeological Inquiry (Washington, DC: Smithsonian Institution Press), pp. 23–64.

Healy, P. F. 1984. 'Northeast Honduras: A Precolumbian Frontier Zone', in F. W. Lange and N. Hammond (eds), *Recent Developments in Isthmian Archaeology: Advances in the Prehistory of Lower Central America; Proceedings of the 44th International Congress of Americanists, Manchester, 1982* (Oxford: British Archaeological Reports), pp. 227–41.

Hendon, J. A. 2003. 'Feasting at Home: Community and House Solidarity among the Maya of Southeastern Mesoamerica', in T. L. Bray (ed.), *The Archaeology and Politics of Food and Feasting in Early States and Empires* (New York: Kluwer), pp. 203–34.

Hodder, I. 1986. *Reading the Past: Current Approaches to Interpretation in Archaeology* (Cambridge: Cambridge University Press).

Jonaitis, A. (ed.). 1991. *Chiefly Feasts: The Enduring Kwakiutl Potlatch* (Seattle: University of Washington Press).

Junker, L. L. 2001. 'The Evolution of Ritual Feasting Systems in Prehispanic Philippine Chiefdoms', in M. Dietler and B. Hayden (eds), *Feasts: Archaeological and Ethnographic Perspectives on Food, Politics, and Power*, Smithsonian Series in Archaeological Inquiry (Washington, DC: Smithsonian Institution Press), pp. 267–310.

Kelly, L. S. 2001. 'A Case of Ritual Feasting at the Cahokia Site', in M. Dietler and B. Hayden (eds), *Feasts: Archaeological and Ethnographic Perspectives on Food, Politics, and Power*, Smithsonian Series in Archaeological Inquiry (Washington, DC: Smithsonian Institution Press), pp. 334–67.

Knight, V. J. 2001. 'Feasting and the Emergence of Platform Mound Ceremonialism in Eastern North America', in M. Dietler and B. Hayden (eds), *Feasts: Archaeological and Ethnographic Perspectives on Food, Politics, and Power*, Smithsonian Series in Archaeological Inquiry (Washington, DC: Smithsonian Institution Press), pp. 311–33.

Lara Pinto, G., and G. Hasemann. 1988. 'La sociedad indígena del Noreste de Honduras en el siglo XVI: Son la etnohistoria y la arqueología contradictorias?', *Yaxkin*, 11.2: 5–28.

Leimgruber, W. 2010. 'Warum feiern wir Feste?', *Uni Nova*, 115: 6–8.

Mainou, L., and A. Aguilar Romero. 2020. 'Proyecto Arqueológico Guadalupe, Colón: Dictamen y propuesta metodológica para la conservación de los restos óseos humanos prehispánicos rescatados en la temporada de campo 2019' (unpublished report).

Mattes, J. 2019. 'Präkolumbische Architektur in Nordost-Honduras' (unpublished bachelor's thesis, Universität Zürich).

McOmish, D. 1996. 'East Chisenbury: Ritual and Rubbish at the British Bronze Age-Iron Age Transition', *Antiquity*, 70.267: 68–76 <https://doi.org/10.1017/S0003598X00082892>.

Piddocke, S. 1965. 'The Potlatch System of the Southern Kwakiutl: A New Perspective', *Southwestern Journal of Anthropology*, 21.3: 244–64.

Reindel, M., P. Fux, and F. Fecher. 2020. 'Archäologisches Projekt Guadalupe: Bericht über die Feldkampagne 2019', *SLSA Jahresbericht*, 2020: 43–59.

———. 2021. 'Archäologisches Projekt Colón, Honduras: Bericht über die Feldkampagne 2020', *SLSA Jahresbericht*, 2021: 33–58.

Sabloff, J. A., and W. L. Rathje (eds). 1975. *Peabody Museum of Archaeology and Ethnology Monographs*, III: *A Study of Changing Pre-Columbian Commercial Systems: The 1972–1973 Seasons at Cozumel, Mexico* (Cambridge, MA: Harvard University Press).

Schaik, C. van, and K. Michel. 2017. *Das Tagebuch der Menschheit: Was die Bibel über unsere Evolution verrät* (Reinbek: Rowohlt).

Smith, S. T. 2003. 'Pharaohs, Feasts, and Foreigners: Cooking, Foodways, and Agency on Ancient Egypt's Southern Frontier', in T. L. Bray (ed.), *The Archaeology and Politics of Food and Feasting in Early States and Empires* (New York: Kluwer), pp. 39–64.

Spinden, H. J. 1925. 'The Chorotegan Culture Area', in *Proceedings of the Twenty-First International Congress of Americanists*, II (Rome), pp. 529–45.

Steadman, L. B., C. T. Palmer, and C. F. Tilley. 1996. 'The Universality of Ancestor Worship', *Ethnology*, 35.1: 63–76 <https://doi.org/10.2307/3774025>.

Stephens, J. L. [1841] 1969. *Incidents of Travel in Central America, Chiapas, and Yucatan* (New York: Harper).

Stone, D. 1941. *Archaeology of the North Coast of Honduras*, Peabody Museum of American Archaeology and Ethnology Memoirs, 9.1 (repr. Cambridge, MA: Peabody Museum).

Strong, W. D. 1935. *Archaeological Investigations in the Bay Islands, Spanish Honduras*, Smithsonian Institution Publication, 3290 (Washington, DC: Smithsonian Institution).

Tozzer, A. 1941. *De Landa's Relación de las cosas de Yucatán*, Papers of the Peabody Museum of American Archaeology and Ethnology, Harvard University, 18 (Cambridge, MA: Peabody Museum).

Urban, P. 1986. 'Precolumbian Settlement in the Naco Valley, Northwestern Honduras', in E. M. Schortman and P. Urban (eds), *The Southeast Maya Periphery* (Austin: University of Texas Press), pp. 275–95.

Vaughn, K. J. 2004. 'Households, Crafts, and Feasting in the Ancient Andes: The Village Context of Early Nasca Craft Consumption', *Latin American Antiquity*, 15.1: 61–88 <https://doi.org/10.2307/4141564>.

JADRANKA GVOZDANOVIĆ

4. Development and Functioning of Numeral Systems

ABSTRACT This paper discusses general properties of counting systems and focuses particularly on the following issues:

a) individuation as a cognitive process presupposed by numerosity,
b) growing and decaying numeral systems,
c) evidence for prehistory from numeral forms and etymologies.

The first issue is general, the second discusses Kiranti Tibeto-Burman languages of Nepal compared to Pama-Nyungan languages of Australia, and the third focuses on (Proto-)Indo-European numerals and evidence on Indo-Uralic and Semitic. The analysis of the decaying Tibeto-Burman numeral systems of Nepal in contact with Indo-Aryan confirms the role of numeral bases and of mathematical operations. It additionally shows that the lexicon and grammar are intertwined: numeral decay usually stops at 'three' and can proceed beyond it only in systems that have reduced the grammatical number beyond 'three' as well. The paper also shows how the analysis of number concepts and numeral forms can be indicative of prehistoric contacts and migrations.

KEYWORDS Number, numeral, base, element, classifier, Tibeto-Burman numeral decay, Indo-European prehistoric evidence

Introduction

Cognition of quantity is generally observed in (higher) living beings, but cognition of *exact* quantity and its numerical evaluation are among specifically human characteristics (cf. a.o. Núñez 2017b). The capacity of counting is at the same time innate and culturally shaped.

Counting starts with individuation of units with distinct boundaries, which thereby become countable. Individuation rests on delimitation and characterization in terms of distinguishable properties. But countability in most instances requires more — an identification as belonging to a set, which can then be cumulatively quantified. Individuation and identification go hand in hand in cognitively construing numerosity. Thus Proto-Indo-European had more than one form for 'one': *H_1oi- (with augments), continued in most Indo-European language groups, *sem/som/sm̥- (in addition to *H_1oi-) continued in Greek and Armenian, and *si-H, continued in Hittite. The root *H_1oi- presumably had the meaning 'single, solitary; one' and *sem/som/sm̥- 'same, self, alone'; *si-H possibly had a deictic meaning of identification. In terms of the discussion presented above, the root *H_1oi- expressed individuation.[1] The other two forms, *sem/som/sm̥- and *si-H, probably expressed identification (and by implicature also individuation). The Proto-Indo-European form *H_1oi- has correlates in Semitic and Berber, Uralic (Samoyed) and Altaic/

1 Interestingly, North-West Caucasian borrowed this form from Proto-Indo-European in the meaning of 'all', implying a chain 'one > every one > all' (Bomhard 2019, 8).

Jadranka Gvozdanović University of Heidelberg. Jadranka Gvozdanović is Professor of Slavic Linguistics at Heidelberg University. She is author or editor of ten books and more than hundred articles, mainly on historical and comparative linguistics. Jadranka Gvozdanović is a corresponding member of the Croatian Academy of Arts and Sciences and winner of the AATSEEL prize for the best book in Slavic linguistics 2009.

Number and Measure in Prehistory: Gift, Exchange, and the Transfer of Goods, ed. by Philippe Della Casa and Aleksander Dzbyński, WSA, 6 (Turnhout, 2024), pp. 57–66.
BREPOLS PUBLISHERS
DOI 10.1484/M.WSA-EB.5.137129

Transeuroasian (Tungus) (cf. Bomhard 2019, 9) and must belong to the beginnings of numerosity (cf. also the last section: 'Numerals Are Indicative of Prehistoric Developments').

Even when counting is based on clearly individuated and identifiable materiality, there is a cognitive evaluation involved. This is made explicit by numeral classifiers occurring in different parts of the world, probably best known from South-East Asia.

Mensural classifiers contribute distinct boundaries and thereby identify an entity (e.g. Burmese *myiʔ tə m̥wa* 'river one section, e.g. a fishing part', Becker 1975, 113; Vittrant 2002). Sortal classifiers attribute characteristics to individuated entities (e.g. Burmese *myiʔ tə 'pa* 'river one sacred object' (e.g. in mythology), Becker 1975, 113; Vittrant 2002). Apparently, individuation and characterization (i.e. identification) result from construals which are at least in part culturally determined. This comes to the fore even more clearly in historical developments of classifiers. According to Bisang (1999, 174), development of classifiers can be either item-oriented or category-oriented. Item-oriented development is exemplified by Old Chinese, where an item had to have a high cultural importance to be worthy of a classifier; Modern Chinese still lacks a specific classifier for [human] or [animate]. In Burmese and Japanese, on the other hand, these categories determine classifiers, and so do shape and function.

Numeral classifiers are instructive concerning the kind and role of materiality in numerosity; they show that all the materiality passes through the cognitive lens of construal (e.g. a river can be counted as a place, a stretched line, an abstract notion, etc.). This construal takes place within the cultural frame, but is in the end decided pragmatically in a communicative setting (e.g. talking about a river as a place for a picnic). In this sense, counting presupposes a situated (or embodied) process of individuation.

Number and Numeral Systems

Individuated entities can participate in sets and be exactly quantified in terms of numbers. Numbers are conceptual units and their linguistic expressions are numerals; together they constitute meaning-form units of a specific kind, because each numeral denotes a number which implies all the lower numbers. Cognitively, numbers are patterned in structures consisting of building blocks, called bases, and elements related to them by means of mathematical operations. Numerals express these structures in accordance with language-specific principles (cf. e.g. *twenty-two* in English vs. *zweiundzwanzig* 'two-and-twenty' in German).

Conceptually, a number base is a building block, a set of elements *n* that can be e.g. used in expressions of the form $xn+y$ (i.e. it can be multiplied by another number and an additional number value can be added to it, cf. also Comrie 2005; subtraction and division occur much more rarely). Prototypical bases are: quinary (5), decimal (10), pure vigesimal (20), decimal-vigesimal (10–20), and hexagesimal (60). Less frequently attested are duodecimal (12) sets used for fractions and multiplication (e.g. for north European big hundreds, cf. Justus 1996). Number bases participate in exponentiation (e.g. 10, 10^{10}) contributing to formation of higher bases; each next-level base is also symbolically conceptualized as higher in ideographic systems of numeral signs (e.g. in ancient Egyptian and Sumerian symbolic counting systems).

Each numeral thus denotes a number (i.e. a cumulative set of countable elements). Numeral symbols representing complex numbers often reveal the mathematical operations that underlie these numbers (e.g. two hundred fifty, i.e. two times hundred and five times ten). Two major principles play a role here: level distinctions among the bases and the principle of linearity unless contradicted by the hierarchical differences, characteristic of bases. The mathematical operations involve in most cases addition, multiplication, exponentiation, and less often subtraction or division. Numerals, expressing numbers, reveal these operations with some degree of transparency; opaqueness is usually caused by sound change blurring morpheme shapes and boundaries.

For example, the Celtic language Breton exhibits far-reaching mathematical operations (cited as number: numeral (explication)):

58: eizh ha hanter-kant (i.e. 8 + ½ 100)
65: pemp ha tri-ugent (i.e. 5 + 3 × 20)
77: seitek ha tri-ugent (i.e. 17 + 3 × 20)
83: tri ha pevar-ugent (i.e. 3 + 4 × 20)
92: daouzek ha pevar-ugent (i.e. 12 + 4 × 20)
108: eizh ha kant (i.e. 8 + 100)
133: kant tri ha tregont / trizek ha c'hwec'h-ugent (i.e. 100, 3 + 30 / 13 + 6 × 20).

Somewhat surprisingly, there is not a specific numeral type of a continent (e.g. quinary, decimal, and vigesimal bases are found across continents). This shows that numeral systems reflect very basic principles of ordering and structure-building that can be assumed to lie at the basis of the human cognition (cf. a.o. Gvozdanović 1999).

The approach presented here agrees with Lakoff and Núñez (2014) and Núñez (2017a and b) in analysing number as exactly quantifying in an abstract and relational way, and referred to symbolically. The essentially relational properties of number also underly Russell's

(1910; 1920) definition of number as cardinality shared by sets of objects.

Symbolic signs representing numbers, i.e. numerals, constitute linguistic subsystems and have often been discussed concerning their patterning (e.g. recently for Pama-Nyungan by Zhou and Bowern 2015, in terms of Bayesian phylogenetics). Numeral systems rest on cognitive properties of the associated number systems, and are intertwined with other grammatical properties of the language in which they occur; the intersection with other meaning-form units makes a study of numeral systems especially revealing concerning underlying systemic and typological properties.

Growing and Decaying Numeral Systems

Analyses of low-level growing numeral systems of hunter-gatherers, such as Pama-Nyungan peoples and languages of Australia, were presented by Zhou and Bowern (2015) in phylogenetic terms. Their analysis yielded the result that 'Paman ancestral state reconstructions suggest two solutions; either limit-three or limit-five. Since reconstructions of limit-three are allocated a higher posterior probability across the entropy range, they are to be preferred' (Zhou and Bowern 2015, 3). Zhou and Bowern also provided evidence for a rapid elaboration of numerals above 'five', in contrast to below 'five', where 'four' is usually innovated compositionally independently of 'three'. All in all, their findings argue against a model of numeral evolution which begins with plain gradual incrementation. What they found in Pama-Nyungan is variation in numeral extent mostly (though not entirely) concentrated around accretion and deletion of numerals between three and five.

Limit-three and limit-five can well coexist within one system whenever they serve different functions, limit-five as a base. In terms of the discussion presented in this paper, and in accordance with the data on Pama-Nyungan languages found by Zhou and Bowern (2015), numeral bases interfere with plain incrementation, while instituting a leap to a higher-order level.

Growing numeral systems are indicative of increased computational needs in a community (e.g. the growth of Sumerian numerals was conditioned by the need for complex astronomical computations); sometimes borrowing occurs. Decaying numeral systems, on the other hand, emerge in multilingual situations, in which the concepts of (particularly higher) numbers come to be expressed by numerals pertaining to the language of a dominant culture.

Language loss, i.e. replacement of indigenous numerals by those of another, as a rule dominating, language provides a countercheck for the data established on growing numeral systems. Such data are pervasively present in Kiranti Tibeto-Burman languages spoken in Nepal.

Kiranti languages are spoken in east Nepal and the north-east of India; the biggest group in east Nepal is the Rai group, and within the Rai group, Bantawa (cf. Fig. 4.1). Kiranti languages are complex pronominalizing (i.e. they mark the person and number of inner arguments of the predicate pronominally in the verb endings) and thereby typologically similar to Pama-Nyungan, but they have (or had, due to decay under the influence of Nepali, an Indo-Aryan language) a more complex numeral system.[2] My investigation in 1984 established different phases of language decay and replacement by the Nepali language. In addition, there was partial replacement of the lost Kiranti forms by language-internal means.

The Rai group is geographically divided into north-eastern, north-western, eastern, western, and central/southern Rai in Nepal. These languages were strongly affected by numeral decay (i.e. replacement of Tibeto-Burman numerals by the corresponding Nepali ones, here indicated as absence of a Tibeto-Burman form) due to total dominance of the official language Nepali. Nepali is the language of mass media and also the language of negotiation in market contexts. This language decay cannot be adequately plotted phylogenetically because a multitude of sociocultural factors play a role, causing different variation in different localities where the same language (Bantawa, Bahing, or Kulung, cf. Table 4.2) is spoken in the neighbourhood of different other languages. The types of variation can be illustrated by the following examples (the data were collected by myself in 1984). For comparison, Table 4.1 starts with Proto-Sino-Tibetan reconstructed forms.

These data illustrate several important points:

- Decay up to 'one' is rare. It can (optionally) occur only if the language system has reduced three grammatical numbers to two (having lost the dual, as in Chourase, cf. also Gvozdanović 1985, 138).
- The only remaining Tibeto-Burman numeral form in Chourase of Dhankuṭa is apparently based on *kol, the Proto-Sino-Tibetan form for 'twenty', representing a numeral base in this language group.
- The form for 'three' goes back to Proto-Tibeto-Burman and has been systematically preserved in the Kiranti languages; the forms for 'one' and

2 The Kiranti languages are pro-drop languages (i.e. they do not obligatorily refer to a missing subject by means of a pronoun), and share in the general SOV word-order typology and modifier-head order of the Indian linguistic area. Functional morphemes (postpositions, subordinating conjunctions, complementizers) follow the elements in their scope (cf. Michailovsky 2018).

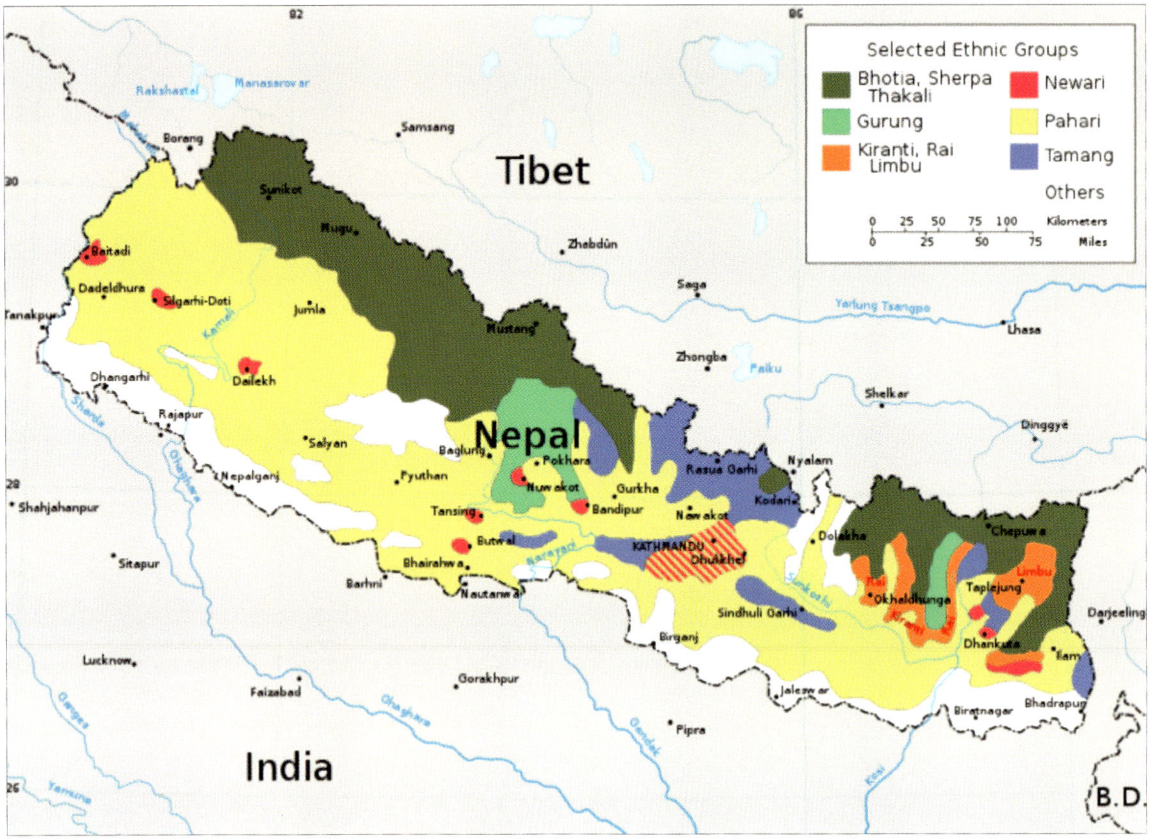

Figure 4.1. Ethnic groups and languages of Nepal (cf. Kiranti in orange) (SVG by Furfur. Original source: CIA work from University of Texas Libraries website: <https://www.lib.utexas.edu/maps/nepal.html>. This file was derived from: Nepal ethnolinguistische Gruppen.svg, <https://upload.wikimedia.org/wikipedia/commons/2/20/Nepal_ethnic_groups.png> [accessed 1 February 2024]).

'two' are language-specific and within a language, constant.

- There is a major cut-off point at 'three' (this resembles the Pama-Nyungan data analysed by Zhou and Bowern 2015).
- The Proto-Tibeto-Burman form for 'five' *ngāH, in (2) illustrated by Bahing of Moli, Kulung of Gudel and Bung, has in Bantawa been replaced by 'one-hand' (and 'two hands' for '10'), patterned on the historical Indo-European model (cf. also section 'Numerals are Indicative of Prehistoric Developments', below).
- Loss of the indigenous form for 'five' opens the way for downward shifts of higher numeral forms up to the boundary of 'three'; in this respect there is also a resemblance with Pama-Nyungan.
- Bantawa exhibits remarkable shifts of *reṭkapok* between 'eight' (by the comparative evidence the original denotation) and 'four'; this shows that the process of decay also follows the mathematical foundations of the number system (i.e. $2 \times 4 = 8$).
- Of the discussed languages, Bantawa is the only one with a supraregional status; its numeral system shows only very few traces of the original Tibeto-Burman system, due to decay and subsequent spontaneous growing of the numeral system by introducing a new base for 'five' meaning 'one hand' and basing the other numerals on it.

Number is part of the lexical system and of the grammatical system (as the grammatical category of number). The numeral decay processes described above show that these two systems in the cognitive area of number are interrelated. This finding concerning the Tibeto-Burman Rai languages might also be investigated concerning Pama-Nyungan, as e.g. Bowern and Zentz (2012, 145) showed that in Worrora, underlyingly the same suffixes are used in the formation of numerals, pronominal number, and verb agreement: singular zero, dual ^{k}*warndu*, trial *oorri*, plural ^{k}*wurri*.

Bantawa and several other north-eastern and eastern languages have the form for 'one' (*ʉk-* in Bantawa, *ek-* in Mewahang, *ik-* in Yakkhaba, cf. Gvozdanović 1985, 136) not based on Proto-Sino-Tibetan, but on Indo-Aryan, attested in the early Middle Aryan treatise on horse training composed by Kikkulis of Mitanni,

Table 4.1. Numerals in Bantawa Rai varieties spoken across Nepal (Gvozdanović 1985, 136–37). Proto-Sino-Tibetan: with some simplifications (reducing variant forms to the arguably more basic one), these reconstructions rest on the data in Matisoff 1997.

	Proto-Sino-Tibetan	Northern Nepal	Central/southern East Nepal	Eastern areas of Nepal	
		Bantawa	Bantawa	Bantawa	Bantawa
		Annapurna	Chhinamakhu	Durdimba	Raniṭar
			Bhojpur	Panchthar	Panchthar
1	*ʔit	ʉkṭai	ʉkṭak/ʉkpok	ʉkbok	ʉkṭa
2	*g-ni-s	hʉapok	hʉapok	hʉsaṭ	hʉaṭa
3	*g-sum	sumkapok	sumkapok	sumka	sumkaṭ
4	*b-liy	reṭkapok	-	-	reṭkaṭaṭ
5	*l-nga	-	-	ʉkchuk	ʉkchuk
6	*d-ruk	-	-	-	bhankachuk
7	*s-nis	-	-	-	bhanhʉchuk
8	*b-r-gya	-	reṭkapok	reṭkabok	bhansumchuk
9	*d-kuw	-	-	-	bhanreṭchuk
10	*gip	-	-	hʉachuk	hʉachuk
20	*m-kul	-	-	-	reṭkachuk

Table 4.2. Numerals in Bahing and Kulung Rai varieties of east Nepal (Gvozdanović 1985, 135).

	Central Nepal	Central Nepal	Central/eastern Nepal	North-eastern areas of Nepal		
	Bahing	Bahing	Chourase	Kulung	Kulung	Kulung
	Moli	Biguṭar	Chhintang	Gudel	Bung	Pawoi
	Okhaldunga	Okhaldunga	Dhankutòa	Solukhumbu	Solukhumbu	Solukhumbu
1	kong	kong	kolo	ibum	ibum	ibim
2	niksi	niksi	-	nicci	nicci	nicci
3	sam	sum	-	supci	supci	sukli
4	le	le	-	lici	lici	likci
5	ngɔ	ruka	-	ngaci	ngaci	tupci
6	rukhu	cani	-	tupci	tupci	tikci
7	cʉni	kodhim	-	-	retci	-
8	ja	-	-	retci	-	-
9	gu	-	-	nuci	-	-
10	kudum	-	-	bɔci	-	-
20	-	-	-	-	-	-

mentioning *aikawartanna* 'one turn of the course' (Sanskrit *ekavartana*). By its linguistic characteristics, this document can probably be dated to the second millennium BC (cf. also Bryant 2001). This coincides with the period of Indo-Aryan migrations into regions including Nepal, which started after 2000 BC (triggered by ecological changes and enabled by the discovery of the war chariot). Judging by the numeral evidence, the Tibeto-Burman inhabitants of east Nepal had an adaptive change to the domineering Indo-Aryan language(s) by adopting the first numeral and preserving some indigenous ones.

Numerals Are Indicative of Prehistoric Developments

Numeral forms and conceptualization of their meanings can offer important insights into historical contacts of peoples including their languages and cultures. As mentioned above, Proto-Indo-European had *H_1oi-, *som/sem/sm̥ — and *si-H, the former probably originally expressed individuation and both latter ones, identification. It comes as no surprise that the root *H_1oi- underlies the forms for 'one' in most later Indo-European languages.

(1) The PIE root *sem/som/sm̥- is continued in the meaning of 'one' in Greek *εἷς* 'one' (Beekes 2010, 394; but Greek also had *οἰνός* 'one (on a dice)', compared to Greek *οἶος* 'sole, single'. *sem/som/sm̥- also underlies Armenian *mi* 'one' (Martirosyan 2010, 467), but Armenian also has *ēn* < *H_1oi- with an *-n-* augment. Tocharian, both A and B, probably offers the clearest example of the numeral 'one' based on the root *sem-, *sm̥- (Adams 2013, 721; Blažek 1999, 152).[3] The other Indo-European languages continue the root *H_1oi- in the meaning of 'one', with different augments (*-kw-*, *-w(a)-*, *-n-*, cf. Blažek 1999) e.g. in Anatolian (cf. Hittite *a-anki*. 'once', *sani-* 'one and the same', cf. Blažek 1999, 142s) and a.o. Indo-Aryan (cf. e.g. Old Indic *éka* 'one' vs. Old Indic *sāmí-* 'half', etc.). Blažek (1999, 156) gives examples from the other Indo-European language groups (cf. Italic, Celtic, Baltic, Slavic, Germanic, Albanian, and Phrygian) and points to similarities of the *H_1oi- root for 'one' to comparable forms in Semitic, Uralic, and Altaic, and for *sem/som/sm̥- in Balto-Fennic and Altaic. Most comparability occurs in Uralic and Altaic (e.g. for *H_1oi-in Samoyedic *oj-*, Korean *oi-*, etc., for *sem/som/sm̥- in Manchu *sonio* 'one, a single', Middle Korean *hannăh*). The Proto-Semitic form is assumed to have been *ašt- 'one', in West Semitic supplanted by *ʔaḥad- 'lone', paralleling the Proto-Indo-European duality in the designation of 'one'.[4]

Thus it would seem that both Proto-Indo-European and Proto-Semitic expressed individuation and identification at the beginning of their counting histories.

(2) 'Two' PIE *dwe/o, with parallels in Altaic (based on the root *tu-*, cf. examples in Blažek 1999, 178s), whereas Semitic is less clear (e.g. Akkadian *šinā*, Aramaic *tərayin tərēn*).[5]

(3) 'Three' PIE *trey-, with no really convincing parallels outside of Indo-European (cf. Akkadian *šalāšum*, Old Aramaic *šlš/lš Ar/*).

(4) 'Four' PIE *kwetur-, with a potential Altaic parallel in *tōr/*tūr 'stretch of four fingers' (Blažek 1999, 215); less clear is a parallel with Akkadian *erbe*, probably borrowed into Old Georgian *rva*.

(5) For 'five', PIE *penk^w-, there are widespread denotations with the same symbolic meaning of 'fist'.

(6) 'Six' PIE *swek̑s 'hand-overgrowing' has clear parallels in Semitic, cf. Akkadian *šeššum*, Aramaic *šett*, Hebrew *šēš* (cf. Lithuanian *šeši*); this is a probable loanword from Semitic.

(7) 'Seven' PIE *septm̥ (Beekes 1995, 215) has clear parallels in Semitic, cf. e.g. Akkadian *sebē*; the Semitic form was borrowed into Kartvelian (*šwid), in line with the numerals 5–10/100 (Manaster Ramer 1995, 16–17); this is most likely a (possibly indirect) PIE loanword from Semitic.

(8) 'Eight' *ok̑to is less clear; if reconstructed as *(H_2)ok̑toH1(u), it may have a parallel in Semitic (e.g. Tuareg (Ahaggar) *ŏkkoẓ*; Blažek 1999, 270).

(9) 'Nine' *neu̯n is usually connected with 'newness'.[6]

(10) PIE *dek̑m̥(t)- 'ten' should according to Szemerényi (1960, 69 n. 69) be analysed as *de-k̑m̥t- 'two hands' (although the form for 'two' should begin with a *dw- (cf. Winter 1992, 17); 'two hands' denote the number 'ten' in many different languages in the world).[7]

3 Saenko (2017) assumes that the root *sem is older for reasons of its omnipresence in derivatives (e.g. Latin *similis*, Old Irish *samail* < *sm̥H_2-el-i 'similar', Gothic *sundro* 'in a single way').

4 For more details of correspondencies cf. Blažek 1999, 156; Beekes 1995, 148.

5 Blažek (1999) mentions parallel derivations in Egyptian, (in his notation) PIE oblique *du-w-oy- (= 'two' + COLLECTIVE + DUAL) = Egyptian *siny-ū-āy (= 'two' + PLURAL + DUAL).

6 Blažek (1999, 287) posits the meaning 'one is in lack', connected with Uralic *ňalV-(kV-).

7 Blažek (1999, 299) hypothesized that it may be related as PIE *dek̑ 'reaching, accomplishing', implying that at the time of the formation this was the highest numeral.

The forms above 'ten' are compounds or periphrases, but 'hundred', PIE *d(e)kṃkómt > *ḱṃtom (Szémerenyi 1960, 139–40) either had the meaning or could be interpreted as the genitive plural of 'hands'. For 'thousand', Blažek (1999, 320) reconstructs three different roots: (1) in the southern Indo-European languages (Blažek refers to Indo-Iranian, Hispano-Celtic, Latin, and Greek) the reconstructed root is *sṃ-ǵ^{h}éslo 'amount (of corn) held in hands', (2) in the north-west, *tuH-sont- based on the root *tuH 'strong' (e.g. Gothic *þūsundi*, Old Church Slavic *tysęšti*), and (3) in Tocharian on *welos 'greatness' attested also in Tocharian.

The striking point about Proto-Indo-European numerals is that 'one' has widespread parallels including Semitic, Uralic, and Altaic, the other lower numerals (except 'three') have direct or indirect parallels in Uralic and Altaic, whereas 'six' and 'seven' are loanwords from Semitic ('eight' might possibly be a direct or indirect loanword from Semitic). This fits well with the recent findings on the prehistoric Indo-Uralic branch related to Altaic, with input from North-Eastern Caucasian (cf. Anthony 2019, Kortlandt 2019).[8] Counting beyond 'five' was probably imported to these originally hunter-gatherer cultures from the agriculturalist Semitic cultures, which needed a more elaborate system for storage, division, and trade of agricultural products. We may assume that counting above 'five' was introduced with the introduction of agriculture, which by the archaeological and genetic evidence spread from the Near East around seven millennia ago (cf. a.o. Cassidy and others 2016). This preceded the dissolution of Proto-Indo-European since the fourth millennium BC (concerning this date cf. Kortlandt 2019).

Here the question arises about loanwords from Proto-Semitic into Proto-Indo-European. Kortlandt (2019, 385) discussed such possible loanwords after eliminating most of correspondences suspected by other authors either for phonological reasons or for reasons of absence from some Indo-European language groups. This led him to list *d^{h}oHn- 'grain', *med^{h}u 'mead', *septm 'seven', and *sweḱs 'six' as probable loanwords from Semitic. These forms are attested in Tocharian and Indo-Aryan, so he concluded that they must have reached Proto-Indo-European by the eastern route. Proto-Indo-European is assumed to have been a branch of Indo-Uralic influenced by archaic North-Eastern Caucasian in the fifth millennium BC and subsequently in the fourth millennium, at the beginning of the disintegration of Proto-Indo-European. Kortlandt assumes that Semitic never directly neighboured Indo-European and loanwords must have come by an indirect way. This of course raises the question of the missing

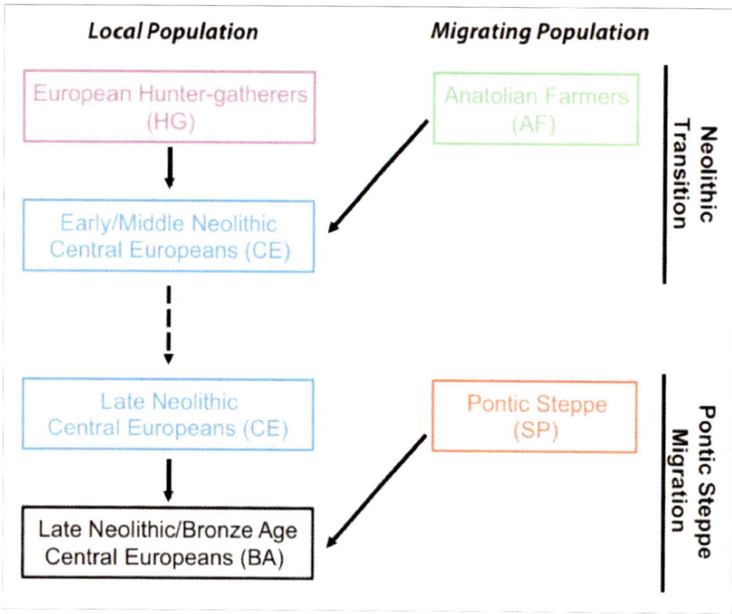

Figure 4.2. Prehistoric migrations into Europe (Goldberg and others 2017, 2658).

attestation of these borrowings in an intermediary (possibly Caucasian) language.

What is striking about Semitic is that its eastern branch, exemplified by Akkadian, exhibits more direct numeral similarities with Indo-European than the other parts. How could these have been transmitted?

In principle there would have existed three possible ways of transmission: (a) by a neighbouring settlement, (b) by trade of agricultural products, or (c) by contact due to migration. Settlement neighbourhood of Semitic to Indo-European, denied by Kortlandt (2019), is still a disputed matter, and archaic texts (which, of course, require cautious reception) make reference to early Semitic inhabitation more to the east (cf. e.g. Peters 1919). This evidence should be treated with caution; moreover, a possibility of indirect influence by way of a Caucasian language cannot be fully ruled out.

Concerning the different possibilities of contact: trade involves traders, and of these we do not have sufficient knowledge. But what we do know, based on archaeogenetic evidence, is that there was a migration from the Near East into Europe.

Goldberg and others (2017) presented specific genetic data on migration from Anatolia in the Neolithic and from the Pontic Steppe into central Europe in the Late Neolithic/Bronze Age (cf. Fig. 4.2). According to their findings, the second of these migrations was performed predominantly by males, whereas there

8 This branch probably dissolved in the seventh millennium BC.

was no sex bias in the preceding migration from the Near East.[9]

Two migration ways were possible and probably applied: across the eastern Mediterranean, or by way of the southern steppe of eastern Europe, encountering steppe herders of eastern Europe and mixing with them. We can hypothesize that either of these ways may have brought Semitic speakers into contact with speakers of Proto-Indo-European during the stage of growing the numeral system.[10]

Other archaeogenetic research reached similar conclusions. Based on archaeogenetic research of sixty-nine Europeans who lived between eight thousand and three thousand years ago, Haak and others (2015) were able to establish that the Yamnaya-culture steppe herders of southern Russia (who built pit graves called *kurgan* known from the Pontic Steppe), migrated into the heartland of Europe around 4500 years ago and contributed around 75 per cent of the Indo-European population of the Late Neolithic Corded Ware people from Germany, descended in part from Near Eastern ancestry. These findings are in agreement with the findings on Pontic Steppe migration by Goldberg and others (2017). Further genetic evidence was offered by Cassidy and others (2016) showing that Ireland was inhabited by two waves of migration, a Neolithic one from the Near East that brought agriculture to Ireland and a Bronze Age migration originating from the Pontic Steppe that introduced metallurgy. It was the second of these migrations that brought Indo-Europeans to Ireland, genetically similar to modern populations. And the first one may in the process have brought Semitic speakers from the Near East into contact with proto-Indo-Europeans before the dispersal.

Such genetic evidence cannot replace cultural evidence, but it can contribute to testing linguistic hypotheses. The hypothesis, based on the numeral shape and meaning, states that the formation of the basic numeral system started in parallel ways in the Near East and the Indo-Uralic and Altaic regions, and the developments became intertwined before the dissolution of the proto-Indo-European unity, assumed to have started in the fourth millennium BC. The archaeogenetic evidence on migrations supports this hypothesis.

Conclusions

The paper has shown that numeral systems represent culturally and cognitively construed numerosity based on *conceived* materiality.

The bases of a number system have clear-cut cognitive values. They are sets having the status of single elements at the next higher level; they are also relatively resistant to change and decay.

Mathematical operations, including addition, multiplication, and exponentiation, in some languages also subtraction and (rarely) division, persist in number systems even if the numeral expressions become opaque or fully absent (as in decaying systems). This is shown by awareness about the mathematical operations even in repairs of decaying systems.

The number systems in grammar and lexicon are interconnected; this becomes obvious in cut-off points for numeral incrementation or decay.

Numeral forms and meanings can provide indications for a historical reconstruction of growing numeral systems and contacts with bearers of other cultures in prehistoric periods. If not refuted by data from other disciplines such as archaeology and archaeogenetics, linguistic evidence can be crucially important for historical reconstruction.

9 The subsequent discussion between Goldman and others and Lascarides and Reich on PNAS did not yield any significant change, cf. <https://doi.org/10.1073/pnas.1704442114>.

10 They were known as the Kurgan peoples and moved into Europe in several waves, mostly in the third and fourth millennia BC (cf. Gimbutas 1979, etc.).

Works Cited

Adams, D. Q. 2013. *A Dictionary of Tocharian B*, 2nd edn (Amsterdam: Rodopi).

Anthony, D. W. 2019. 'Archaeology, Genetics, and Language in the Steppes: A Comment on Bomhard', *Journal of Indo-European Studies*, 47: 175–98.

Becker, A. L. 1975. 'A Linguistic Image of Nature: The Burmese Numerative Classifier System', *Linguistics*, 13.165: 109–22.

Beekes, R. S. P. 1995. *Comparative Indo-European Linguistics: An Introduction* (Amsterdam: Benjamins).

———. 2010. *Etymological Dictionary of Greek* (Leiden: Brill).

Bisang, W. 1999. 'Classifiers in East and Southeast Asian Languages. Counting and Beyond', in G. Jadranka (ed.), *Numeral Types and Changes Worldwide* (Berlin: De Gruyter), pp. 118–85.

Blažek, V. 1999. *Numerals* (Brno: Univerzita Masarykova).

Bomhard, Allan R. 2019. 'Additional Proto-Indo-European/Northwest Caucasian Lexical Parallels: Supplement 1' <https://www.academia.edu/43240276/Bomhard_Additional_Proto_Indo_European_Northwest_Caucasian_Lexical_Parallels_Supplement_1_2020_> [accessed 12 June 2022].

Bowern, C., and J. Zentz. 2012. 'Diversity in the Numeral Systems of Australian Languages', *Anthropological Linguistics*, 54.2: 133–60.

Bryant, Edwin. 2001. 'Myths of Origin: Europe and the Aryan Homeland Quest', in Edwin Bryant, *The Quest for the Origins of Vedic Culture: The Indo-Aryan Migration Debate* (Oxford: Oxford University Press), pp. 13–45.

Cassidy, L. M., R. Martiniano, E. M. Murphy, M. D. Teasdale, J. Mallory, B. Hartwell, and D. G. Bradley. 2016. 'Neolithic and Bronze Age Migration to Ireland and Establishment of the Insular Atlantic Genome', *Proceedings of the National Academy of Sciences*, 113.2: 368–73.

Comrie, B. 2005. 'Numeral Bases', in M. Haspelmath, M. S. Dryer, D. Gil, and B. Comrie (eds), *The World Atlas of Language Structures* (Oxford: Oxford University Press), pp. 530–33.

Gimbutas, M. 1979. 'Proto-Indo-European Culture: The Kurgan Culture during the Fifth, Fourth and Third Millennia B.C.', in G. Cardona, H. M. Hoenigswald, and A. Senn (eds), *Indo-European and Indo-Europeans* (Philadelphia: University of Pennsylvania Press), pp. 155–97.

Goldberg, A., T. Günther, N. A. Rosenberg, and M. Jakobsson. 2017. 'Ancient X Chromosomes Reveal Contrasting Sex Bias in Neolithic and Bronze Age Eurasian Migrations', *Proceedings of the National Academy of Sciences*, 114.10: 2657–62.

Gvozdanović, J. 1985. *Language System and its Change* (Berlin: De Gruyter).

———. (ed.). 1999. *Numeral Types and Changes Worldwide* (Berlin: De Gruyter).

Haak, W., I. Lazaridis, N. Patterson, N. Rohland, S. Mallick, B. Llamas, G. Brandt, S. Nordenfelt, E. Harney, K. Stewardson, Q. Fu, A. Mittnik, E. Bánffy, C. Economou, M. Francken, S. Friederich, R. G. Pena, F. Hallgren, V. Khartanovich, A. Khokhlov, M. Kunst, P. Kuznetsov, H. Meller, O. Mochalov, V. Moiseyev, N. Nicklisch, S. L. Pichler, R. Risch, M. A. R. Guerra, C. Roth, A. Szécsényi-Nagy, J. Wahl, M. Meyer, J. Krause, D. Brown, D. Anthony, A. Cooper, K. Werner Alt, and D. Reich. 2015. 'Massive Migration from the Steppe Was a Source for Indo-European Languages in Europe', *Nature*, 522: 207–11.

Justus, C. 1996. 'Numeracy and the Germanic Upper Decades', *Journal of Indo-European Studies*, 24: 45–80.

Kortlandt, F. H. H. 2019. 'Indo-European between Uralic and Caucasian', *Journal of Indo-European Studies*, 47: 383–88.

Lakoff, G., and R. Núñez. 2014. *Where Mathematics Comes from* (Tokyo: Maruzen).

Manaster Ramer, A. 1995. 'Some Borrowed Numerals in Proto-Kartvelian', *Dhumbadji*, 2–3: 16–18.

Martirosyan, H. K. 2010. *Etymological Dictionary of the Armenian Inherited Lexicon* (Leiden: Brill).

Matisoff, James. 1997. *Sino-Tibetan Numeral Systems: Prefixes, Protoforms and Problems*. Pacific Linguistics, Series B, volume 114 (Canberra: The Australian National University).

Michailovsky, B. 2017. *Kiranti Languages* <https://halshs.archives-ouvertes.fr/halshs-01705023> [accessed 12 June 2022].

Núñez, R. 2017a. 'Biological Enculturation beyond Natural Selection', *Trends in Cognitive Sciences*, 21.6: 404–05.

———. 2017b. 'Is There Really an Evolved Capacity for Number?', *Trends in Cognitive Sciences*, 21.6: 409–24.

Peters, J. 1919. 'The Home of the Semites', *Journal of the American Semitic Society*, 39: 243–60.

Russell, B. 1910. 'The Theory of Logical Types', in D. Lackey (ed.), *Essays in Analysis* (New York: Braziller), pp. 215–52.

———. 1920. *Introduction to Mathematical Philosophy*, 2nd edn (London: Allen and Unwin).

Saenko, M. N. 2017. 'Istorija semantiki praslavjanskix leksem *edinŭ i *samŭ' [History of the Semantics of the Proto-Slavic lexemes *edinŭ and Üsamŭ], *Slověne*, 2017: 76–93.

Szémerenyi, O. 1960. *Studies in the Indo-European System of Numerals* (Heidelberg: Winter).

Vittrant, A. 2002. 'Classifier Systems and Noun Categorization Devices in Burmese', *Proceedings of the Annual Meeting of the Berkeley Linguistics Society*, 28.1, special issue: *Special Session on Tibeto-Burman and Southeast Asian Linguistics*: 129–48.

Wilson-Wright, A. 2014. 'The Word for "One" in Proto-Semitic', *Journal of Semitic Studies*, 59.1: 1–13.

Winter, W. 1992. 'Some Thoughts about Indo-European Numerals', in J. Gvozdanović (ed.), *Indo-European Numerals* (Berlin: De Gruyter), pp. 11–28.

Zhou, K., and C. Bowern. 2015. 'Quantifying Uncertainty in the Phylogenetics of Australian Numeral Systems', *Proceedings of the Royal Society*, 282: 20151278 <http://dx.doi.org/10.1098/rspb.2015.1278>.

JÜRG HELBLING

5. Some Further Thoughts on Commodity and Gift Exchange in Tribal Societies

ABSTRACT Some central aspects and theories of gift exchange and commodity exchange in tribal societies are discussed. It is argued that through ceremonial gift exchange, political relations (military alliances, marriage contracts) are established, reinforced, or repaired, and politically relevant status positions (big man, chief) are attained and validated. In contrast, commodity exchange, which in simple societies often takes the form of barter trade, is an economic activity in which goods are exchanged at a certain price and for mutual benefit of the exchange partners. Gift exchange and commodity exchange exist simultaneously, with commodities often entering a society through long-distance trade, being monopolized by the older men, and transformed into gifts that circulate between the local groups of a society.

KEYWORDS Gift exchange, commodity exchange, barter, military alliance, status competition, political interest

Gift exchange and commodity exchange have been hotly debated in anthropology since the publication of *The Gift* by Marcel Mauss. Not all theories will be addressed here, nor will I deal with gift exchange in our societies. Instead, I will focus on some essential aspects of gift exchange and commodity exchange in tribal societies.[1] I will do this in eight points.

'Archaic Gift Economies' and Commodity Exchange

The distinction between commodity exchange and gift exchange is notoriously ambivalent and vague in Mauss (1923–1924), but nevertheless always present. Like Weiner (1985; 1992), Strathern (1988; 1992), and Graeber (2001; 2011), Mauss thinks that commodity exchange is only practised in complex societies, whereas gift exchange dominates all exchanges of goods and services in simple societies. In these so-called 'archaic gift economies' — or societies of a 'non-calculating family-communism' (Graeber 2011) — there is no separate sphere of commodity exchange.

1 A tribal society is a regional population consisting of politically autonomous local groups (villages) of farmers, sedentary fishermen, and/or pastoralists who are not yet or no longer controlled by a state (cf. Haas 1990, 172; Bailey 1969, 12). For this reason, wars can break out between villages at any time (Helbling 2006, 116–50). Local groups form alliances against common enemies, which are strengthened by gift exchange and marriage relations (Sahlins 1968). The concept of tribal society includes 'simple chiefdoms', where the position of a chief does not exceed the authority and power of local elders and big-men. Basically, a chief is the leader of an enduring alliance between the villages of his domain. ('Complex chiefdoms' often are a kind of state or proto-state, where a chief and his lineage dominate, with the help of a following of kinsmen and war-slaves.)

Jürg Helbling University of Lucerne. Jürg Helbling was Professor for Social Anthropology at the University of Zürich and Lucerne. He is currently Senior Professor for Economic and Political Anthropology at the University of Lucerne. He has conducted fieldwork in the Philippines. His main fields of interest are economic and political anthropology. He has published *Neue Politische Ökonomie einfacher Gesellschaften* (Wiesbaden: Springer, 2021) and *Tribale Kriege: Konflikte in Gesellschaften ohne Zentralgewalt* (Frankfurt: Campus, 2006), and has edited together with Tobias Schwoerer *The Ending of Tribal Wars: Configurations and Processes of Pacification* (London: Routledge, 2021).

Number and Measure in Prehistory: Gift, Exchange, and the Transfer of Goods, ed. by Philippe Della Casa and Aleksander Dzbyński, WSA, 6 (Turnhout, 2024), pp. 67–77.
BREPOLS PUBLISHERS DOI 10.1484/M.WSA-EB.5.137130

However, this view is a result of an analytical shortcoming rather than an ontological fact. Chapman (1980), Gell (1992), Godelier (1977; 1996), Görlich (1992; 1996), Gregory (1997, 47), but also Malinowski (1922), Bohannan (1955; 1959), Barth (1967), and many others have shown that both gift exchange and commodity exchange do in fact exist simultaneously in tribal societies. Moreover, these two modes of transaction are conceptually distinguished even by the people themselves, as on Trobriand (*kula* and *gimwali*, *wasi* and *vava*), among the Maring (*munggoi awom* and *munggoi rigima*), and the Tiv (*shagba* and *yiagh*) to name but a few.

Commodity exchange and gift exchange have been associated with a variety of conceptual distinctions. The difference between commodity exchange and gift exchange is said to correspond to the difference between: interest-led and non-interest-led action (Mauss 1923–1924) or rational and irrational behaviour (Berthoud and Sabelli 1976) respectively, egoism and altruism (Sahlins 1965), liquidating and non-liquidating (Znoj 1995), legally reclaimable and legally non-reclaimable transactions (Testart 2001), immediate and delayed return (Sahlins 1965), as well as alienable and inalienable objects (Mauss 1923–1924; Weiner 1985; 1992). I will deal with these conceptual dichotomies in the following and explain why I think they are not appropriate. But first, what is commodity exchange and what is gift exchange?

Defining the Two Modes of Exchange

Based on a variety of ethnographic examples and inspired by Gregory (1982, 13, 39, 41) — although not fully aligned with him — I propose the following definitions.[2]

Commodity exchange is an economic exchange which is based on a division of labour (i.e. specialization) and comes about through a quantitative ratio (i.e. a price) between the goods exchanged. Gift exchange, on the other hand, is a political exchange that establishes, affirms, or restores socio-political relations between individuals and groups and, thus, defines a qualitative relation between them. Gifts — unlike commodities — usually circulate in ceremonial-ritual contexts. But in both categories, a transaction will only take place if it is in the interest of both parties, usually belonging to different local communities or local kin groups respectively.

It is also true that one and the same object can serve once as a gift and another time as a commodity (Kopytoff 1986): a pig can be exchanged for another commodity (barter) or sold on the market for money, it may also be given away as a gift, distributed by a big-man (redistribution), donated to affinal relatives or to allies during a feast, or it may be consumed in a household.[3] However, the difference between political gift exchange and economic commodity exchange remains clear, as does the different logic according to which objects circulate in the two different spheres.

In the course of this text, I will try to substantiate this definition of these two types of exchange and illustrate it with ethnographic examples.

Value and Price

The distinction between commodity exchange and gift exchange corresponds to two different concepts of value. Value in the sense of exchange value is the price of goods. Godelier (1977), Gewertz (1978), Malinowski (1922), Modjeska (1985), and others have analysed exchange rates (i.e. prices) in commodity exchange in various tribal societies. Exchange rates are obviously the result of a bargaining process: if two exchange actors had not agreed on an exchange rate, the transfer would not have taken place. A successful transfer of commodities is always in the mutual interest, even it is an unequal exchange by the standards of the labour theory of exchange value. Godelier (1977) has shown that this theory of exchange value does not sufficiently explain the price for commodities in tribal societies, although the labour input plays a role. It seems that the actual price of a transferred commodity lies between a minimum exchange rate, determined by the seller's production costs, and a maximum exchange rate,

2 'Gift exchange is an exchange of inalienable goods between individuals between whom relationships already exist; gift exchange establishes a qualitative relationship between the exchange partners. Commodity exchange, on the other hand, is an exchange of alienable goods between individuals between whom no relationships exist; market exchange establishes a quantitative relationship between the goods' (Gregory 1982, 13). 'Commodity exchange establishes objective quantitative relationships between the objects transacted, while gift exchange establishes personal qualitative relationships between the subjects transacting' (Gregory 1982, 39). 'Commodities are alienable objects transacted by aliens; gifts are inalienable objects transacted by non-aliens' (Gregory 1982, 41). The exchange of commodities does not only and not primarily take place between strangers, and the exchange of gifts often serves to establish relationships with people who previously had no relationship with each other.

3 Even in tribal societies money (in the form of coins and notes) is sometimes used as a gift as in our own society (cf. Healey 1985 on the Maring).

depending on the buyer's intensity of demand for the commodity in question. This may be illustrated with the following example. In the Sepik region, Chambri and Garamambu exchange fish for sago. Chambri women sell fish at a minimum exchange rate (which is the maximum rate for Garamambu women who sell sago) of four fish for three units of sago. For Garamambu women, the minimum exchange rate (and maximum rate for Chambri women) is two fish for five units of sago. The exchange rate will lie between 4 F = 3 S and 2 F = 5 S (cf. Modjeska 1985, 159–60). If both trading partners have about the same bargaining power, the price could well lie in the middle, i.e. at three fish = four units of sago. However, the Garamambu are militarily inferior to the Chambri (and the Iatmul, allies of the Chambri). This asymmetry leads the Garamambu to having to accept an unfavourable exchange rate of two fish for five units of sago, even when sago gets scarcer, as in the case of the neighbouring Sawo (Gewertz 1983, 31; 1978, 39, 79–80, 139).

Value in gift exchange, in contrast, has a different meaning. In gift exchange, 'value' refers to the 'esteem' and the 'importance' of a relation for the exchange partners, expressed by means of gifts (Gregory 1982, 47–51; Strathern 1992). This value is determined not only qualitatively (the preciousness of a gift) but also quantitatively (the number of gifts given). Gifts are usually ranked in categories: among the Mai Enga, pigs and cassowaries are the most valuable gifts, followed by pearl-shell pendants, pork sides, stone axes, and cassowary feather ornaments; then two categories of several kinds of shells, plumes, and pelts and the lowest category, consisting of vegetable foods consumed during ceremonies (Meggitt 1974, 170). The quantity of gifts donated counts because the appreciation of an exchange partner is also expressed by the quantity of the exchanged gifts: if too few (of a certain kind of) gifts are donated, the exchange partner will feel held in low esteem. If too high a quantity of gifts is given, the recipient is put under pressure and risks a decrease of his status in relation to the donor in case he is unable to reciprocate. In general, the relationship between the two exchange partners is negotiated through the relative quantity and quality of gifts as well as the timing of their exchange.

'Inalienabilty' and 'Liquidability' in Gift Exchange?

Gregory (1982, 13, 41) agrees with Mauss and Weiner that gifts are ultimately — and unlike commodities — 'inalienable'. According to these authors, this 'inalienability' explains the reciprocation of a gift: the spirit of the donor (*hao* in the case of the Maori) somehow adheres to the gift and compels the recipient to reciprocate.[4] In tribal societies, however, only (collectively owned) land, valuable heirlooms, and (usually) ritual positions may not be exchanged and are truly 'inalienable'. But gifts are in fact alienable, since 'in making a prestation, the donor loses access to the exchange object', as Gell (1992, 145) has pointed out.

Other authors consider gift transactions — in contrast to commodity transactions — to be non-liquidable (Gregory 1982; Znoj 1995). It is through the delayed reciprocation of a gift that an obligation is created between the exchange partners which is intentional and wanted. In commodity exchange, too, there can be a delayed repayment in the form of credit. But in this case, debt is a burden since the credit has to be paid back with interest.[5]

The question, however, is whether a gift obligation is not liquidated by the exchange partner reciprocating the gift. According to Gregory (1982, 53) gift exchange is propelled by a desire to avoid an evenness in exchange balance and to maintain a difference in obligation, i.e. every transactor must give more than he has received. Thus: 1 gives a, 2 reciprocates a+b, 1 then donates a+b+c, and 2 reciprocates a+b+c+d, etc. (cf. Strathern 1971, 216 on *moka* among the Melpa). However, there is no such inflationary ratchet in ethnographic reality, except for the *potlatch* among the Kwakiutl.[6] Rather, gift transactions are liquidating (though by a time delay) too, as in *kula* exchange on Trobriand: one *kula*-actor (A) gives a shell armband (*mwali*) as a gift (*vaga*) and expects a red shell necklace (*soulava*) in return (*yotile*) a year or two later. It may be that his *kula*-partner (B) has no suitable necklace of comparable value. In that case, B will give a small necklace as an interim gift (*basi*) to show his willingness to reciprocate with a necklace of comparable value. As soon as he has such a necklace, B will give it to A, who will give him a small armband in

4 Firth (1959, 419), the main ethnographer of the Maori, criticizes Mauss for having misunderstood the concept of *hao*. In the eyes of the Maori, the *hau* does not contain any active qualities. There is no identification of giver and gift. A man who does not return a gift stupidly terminates a relationship, he may have been interested in, and may even be bewitched for not reciprocating a gift (Firth 1959, 412, 421).

5 However, there is not always an interest in being paid (cf. Maring), but Mai Enga men who participate in *te*-exchange pay those not participating for rearing pigs for them (Meggitt 1974).

6 The actors in the *potlatch* practice such a mutual outdoing by means of gift giving, but they do not seek to maintain the mutual shifting debt differential, but to establish a unilateral debt differential in order to defeat the rival and put him in an inferior status.

Table 5.1. Gift exchange among the Melpa.

I			
A (x gifts)	→ B	B (x + y gifts)	→ A
II			
B (x gifts)	→ A	A (x + y gifts)	→ B

Table 5.2. Gift exchange among the Mai Enga.

I		
Initiatory gifts (*saandi*)	Main gifts (*te pingi*)	Counter-gifts (*yae pingi*)
A → B → C → D	A ← B ← C ← D	A → B → C → D
II		
Initiatory gifts (*saandi*)	Main gifts (*te pingi*)	Counter-gifts (*yae pingi*)
A ← B ← C ← D	A → B → C → D	A ← B ← C ← D

return for his *basi* (Malinowski 1922). So *basi* is clearly not a credit, as Testart claims (Darmangeat 2016, 37).[7]

Even in *moka*-exchange among the Melpa there is an 'equilibrium in the longer run', i.e. after two subcycles (I and II each consisting of two transactions each), through 'alternating disequilibrium' within each subcycle (Strathern 1971, 11) (Table 5.1).[8]

The Mai Enga are a similar case in point except for that their *te*-exchange is a system of two subcycles (I and II) consisting of three transactions each (Table 5.2).

These three different steps in gift transactions are clearly part of gift exchange and have nothing to do with commodity exchange, as Darmangeat (2016, 37–40) maintains, although Mai Enga men who participate in *te*-exchange pay those not participating for rearing pigs for them, which is a payment with interest (Wiessner and Tumu 1998).

Let us now turn back to commodity exchange.

Commodity Exchange, Money, and Two Misunderstandings

Commodity exchange — 'commercial transaction', 'trade', or 'market transaction' (Dalton 1965) — is an economic exchange which is based on a division of labour and comes about through a quantitative ratio (i.e. the price) between the commodities exchanged. Commodities include all goods that are sold and bought: either through barter (C → C, 'moneyless market exchange') or for money (C → M → C; Dalton 1965, 49, 52).[9]

We briefly have to speak about money.

Money

Money is a commodity that can be exchanged for a wide range of other commodities. Money has three functions: a) money as a means of payment (medium of exchange) for exchange against a variety of other goods; b) money as a standard of value, i.e. a unit of account for fixing rates of exchange; and c) money as a store of value, i.e. as a saved means of payment for future exchange transactions (Polanyi 1957, 264–66; Schneider 1974, 158–59, 162–64).[10]

A 'money commodity' must have certain properties so that it can fulfil its function as a means of payment, namely: durability (for storage), divisibility (for small and large payments), manageability (good transportability), and liquidity (availability in sufficient quantities). If these properties are given, any commodity can function as money: such as shells, coins, precious metals, animals, dogs' teeth, cigarettes, but also salt, tea bricks, or sugar (Schneider 1974, 163–64, 172). The only decisive factor is whether it is accepted as a means of payment by the other market participants.

Mauss (1923–1924, 130), Sahlins (1974), and Hart (1986, 649; 2005, 160, 172) consider gifts to be a kind of money with 'purchasing power' or as a means of payment for the repayment of debts/obligations and the acquisition of rights (to memberships, to wives

7 Actors in *kula* exchange also try to obtain objects of greater value than those they have given. But this does not correspond to the logic of the *kula*, but is the result of the individual shrewdness and cunning of *kula*-actors.

8 Moreover, according to Rospabé (1993; 1995) gifts are not payments anyway, but merely the recognition of a debt to return a life: for a woman received or for a death caused. That means that in many gift transactions — at least as far as marriage exchange and compensation is concerned — exchange partners are even if 'a life has given back'. That is why among the Nuer, even after the payment of blood wealth, the debt is not (yet) liquidated (Evans-Pritchard 1940).

9 Dalton (1965, 46–48) distinguishes between 'commercial transactions' (commodity exchange, market exchange, including barter) and 'non-commercial transactions' (reciprocity and redistribution). In capitalist commodity economies, where the market is the dominant mode of transaction, all-purpose money (dollars, francs, or euros) is used in all transactions: not only in commercial exchange (of commodities), but even in non-commercial transactions: reciprocity and redistribution.

10 The first function (standard of value and unit of account) is a precondition for the second function (means of payment and exchange), while the reverse is not true. I will use the term 'money' only when it is used as a means of payment and exchange.

Table 5.3. Commodity trade of the Tiv.

Tiv	trade with	**Hausa (in the north-west)**
tugudu cloth, slaves,	for	cattle, horses, brass rods,
ivory, kola-nuts	cloths, dyed textiles, glass beads	
Tiv	trade with	**Igbo (in the south-west)**
brass rods, cattle, horses	for	muskets, gunpowder
Tiv	trade with	**Fulbe (in the north-east)**
cattle	for	women

and children), but also occasionally for the purchase of goods. I follow Malinowski (1921, 13–14), Sillitoe (1978), Gregory (1982), Strathern and Stewart (2005), and others who not only distinguish between gifts and commodities but also assign money solely to the exchange of commodities. Goods can be transformed into gifts, but may also be used as money (in order to buy other goods in barter trade), but gifts are not money, the only exception being when families in times of hunger use gifts to buy food.

In simple commodity exchange only goods can be bought for money, not labour nor land. In general commodity exchange (almost) all goods, including labour and land, can be sold and bought for money (Bohannan and Dalton 1962; Gregory 1994). Polanyi (1957, 264–66) and Dalton (1965, 51–52) distinguish between a) 'special purpose equivalent', b) 'general equivalent' (for most goods except labour and land), and c) 'universal or all-purpose equivalent' (cf. also Chapman 1980, 54). However, not everything can be bought even with all-purpose money: even in a capitalist commodity economy, not all goods can or may be freely sold and bought, such as human organs, drugs, people, as well as sports trophies, academic titles, and political offices (Bloch and Parry 1989, 13).

Barter: A Myth?

Whereas Polanyi (1957), Chapman (1980), Gell (1992), Godelier (1977; 1996), Görlich (1992), and Healey (1984; 1985), and others have shown that barter between local communities does indeed exist in tribal societies (and even among Australian Aborigines, McCarthy 1938–1939; 1939–1940), some theorists consider barter a myth (among them Strathern 1988; Graeber 2001; 2011; Heady 2005).[11] Two reasons may be responsible for such a misunderstanding.

First, in tribal societies not only goods with an economic use, i.e. food and tools, are traded, but also objects which later are used as gifts, i.e. for political and symbolic purposes: shells, pigs, and feathers among the Maring and Mai Enga (Healey 1984; Meggitt 1974) or brass rods and *tugudu* cloth among the Tiv (Bohannan 1955). Usually, local elite persons (chiefs of local lineages, villages, or of a domain) are the 'relay stations' where this conversion of imported commodities into gifts takes place, as Douglas and Isherwood (1996) have shown for the Tiv (Table 5.3).

Such a conversion of commodities into gifts (C → G) also takes place when a Trobriand chief pays craftsmen with goods (food and meat) for producing *kitoum* objects which he then uses as gifts (*vaygua*) in *kula* transactions (Singh Uberoi 1971). Thus, the local elites monopolize and control the gift exchange.

There are reverse cases of gifts being exchanged for food (G → C), for instance if a family in times of need is forced to sell off its heirlooms and valuables — brass rods among the Tiv (Bohannan 1955), bronze gongs and Chinese jars among the Iban (Freeman 1970) — to buy food. These transactions, however, are considered highly undesirable because they entail a loss of status.

In addition, and this is the second reason for the misunderstanding, commodity exchange between individuals from different villages — especially if it is based on credit — usually takes place only if there is already a social relationship of kinship and marriage, of

11 Although Smith ([1776] 1999) mentions a 'natural propensity to trade' — and Graeber focuses primarily on this idea — there is an alternative theory of trade in Smith's work, which he in fact pursues in his book: people want to increase the quantity and diversity of goods, i.e. their productivity. To this end, they specialize in the production of those goods that they can produce particularly efficiently (or that are available to them). This leads to a division of labour, which at the same time entails the exchange of goods.

Table 5.4. Commodity exchange between Kuboma, Sinaketa and Dobu.

Phase 1		Phase 2	
Sinaketa →	← Kuboma	Sinaketa →	← North-west Dobu
12 coconuts	1 *tanepopo* basket	1 *tanepopo* basket	12 coconuts + 1 unit of sago + 1 belt
4 coconuts	1 comb	1 comb	4 coconuts + 1 bundle of betel
8 coconuts	1 armband	1 armband	8 coconuts + 2 bundles of betel
12 coconuts	1 chalk container	1 chalk container	12 coconuts + 2 units of sago

gift exchange and/or alliance between the trade partners. Gift exchange is thus a precondition for a more or less smooth commodity exchange (Singh Uberoi 1971 on Trobriand), as are kinship and marriage (Healey 1984; 1985 on the Maring) or personal relationships of an elder man with long-distance traders (Guyer 2004 on the Tiv). However, the goal of commodity transactions remains an economic exchange of goods based on a division of labour and exchanged at a certain price (Modjeska 1985).

The exchange of fish for tubers between inland and coastal villages on Trobriand is a simple, direct trade (*vava*) between households, provided that social relations between the villages are in a good condition. If relations between the two villages have been damaged by conflicts, they have to be repaired. This is achieved by the chiefs exchanging valuables (*wasi*), thereby creating the political conditions for a resumption of peaceful trade between the households. The gifts exchanged constitute the pledge of the chiefs' personal responsibility to ensure peace between the groups involved (Malinowski 1922, 187–90; 1935, 17, 42–43).

Likewise, the *gimwali*-trade takes place between *kula*-districts. Men from Sinaketa visit their *kula*-partners (*karayta'u*) in north-west Dobu every two years. Before leaving for north-west Dobu, the people from Sinaketa visit the village of Kuboma to buy trade goods, which they will sell in north-west Dobu, with profit as table 5.4 shows.

People of two *kula*-districts (as Sinaketa and north-west Dobu) who exchange gifts also trade with each other. However, a man may not trade with his direct *kula*-partner (*karayta'u*), but only with the other members of his *kula*-partner's district (Malinowski 1922, 363).

At the same time, gift exchange remains a political transaction, establishing, reaffirming, or repairing socio-political relations (of marriage or alliance). To summarize with Dalton: 'Commercial transactions' are 'embedded' in social relations, but nevertheless remain economic transactions. 'Non-commercial transactions' (gift and marriage exchange) are political transactions that create relationships of trust, which come into play in the context of military alliances, but also as preconditions for a smooth commodity exchange, especially if credit becomes involved.

Indeed, it is in barter where credit is important because of the problem of double coincidence. In a simple one-to-one transaction, trade partner A must find another (B) who wants to give the good that A wants and in return wants to have the good that A wants to give away, at the same time. However, suppose A has a bundle of feathers and wants an axe in return, B has an axe and but wants a fur in return. If the trade is to take place, there are three possibilities: 1) trade partner A must find a third trade partner (C) who wants to give a fur in order to get feathers. A makes this trade and then pays B for an axe with a fur. This may take time. And the question is whether trade partner B will still be willing to barter then. 2) Trade partner B can give an axe to A on credit, and hopes that the latter will pay him sooner or later with a fur. But in this case the problem arises whether trade partner A will keep his promise. 3) Trade partner B may accept feathers from A in exchange for an axe; he then will exchange the feathers for a fur with C, provided that C has not yet carried out his transaction with another trade partner (D, E, or F).

Commodity exchange is always and everywhere 'embedded' as various schools of institutionalism (from Polanyi and Dalton to Coase, North, and Williamson) have demonstrated. The question however is how. In tribal societies, personal relations of kinship and marriage and of gift exchange exist between trading partners, and there are institutions that informally regulate commodity transactions by conventions.

Let us turn back to gift exchange again.

Gift Exchange, Alliance, and Status Competition

Gift exchange — as mentioned above — is about establishing, reaffirming, or restoring socio-political relations between individuals and groups. Gift exchange is, thus, a political exchange that establishes and reaffirms symmetrical relations between individuals and groups (marriage relations, political alliances) or restores them, if damaged (gifts of compensation).[12] In addition, gift exchange also creates and validates asymmetrical relationships or it serves as a medium for status rivalry.

Thus, gift exchange serves: a) the establishment of alliance and marriage relations, b) as compensation payments after wars or in the case of legal disputes, and c) the acquisition and validation of higher status positions.

Gift Exchange and Alliance

In tribal societies, wars between villages are common and frequent (Helbling 2006; Keeley 1996; Sahlins 1968). Those who have to wage war need allies against common enemies, but there is always mistrust between allies: suspicion of poor support, of unreliability, or even betrayal by secretly making a separate peace deal with the enemy. No local group knows for sure to what extent another group is interested in an alliance, since alliance partners usually have alternative alliance options. Hence, trust — established by gift exchange — is instrumental and decisive.

A group signals to another group its interest in an alliance with gifts and commits itself by using them as a pledge (Camerer 1988). The gifts must be sufficiently valuable (i.e. costly), because this is the measure for the interest and commitment in an alliance with the exchange partner (Gintis and others 2001; Smith and Bliege 2005). The quantitative aspect also plays a role as well. Therefore, gift transactions are always collective affairs (contributions from all members of a kin group), and it takes time to raise the gifts in sufficient quantities, and the reciprocation of gifts often takes several years. The gifts signal the extent of interest in friendly relations and stand for the sincerity of one's intention to establish and maintain them.

The exchange of gifts must be balanced over time, because an imbalance in the exchange (gifts not in an appropriate quantity and quality) would be seen by the exchange partner as a sign of disrespect and create mistrust and suspicion. The alliance would be terminated if the gifts were not reciprocated at the appropriate level. Furthermore, the gifts must not be reciprocated too early, since what counts is the mutual obligation created by gift exchange. However, one should not wait too long in reciprocating a gift either, because this may cause the suspicion that one has no intention to reciprocate it at all. Not surprisingly, gift exchange often leads to conflict as it is the medium in which the shaping of a relation is negotiated (Fortune 1932 on *kula*; Pospisil 1994 on the Ekagi-me).

The mutual obligation, created by gift exchange, is used to build up a political relationship: not only alliances, but also status differences.

Gift Exchange and Status Competition

Gift exchange can also be used to create and validate asymmetrical relationships or status differences. The transaction of gifts is usually controlled by the local power elites, i.e. older men, big-men, or chiefs (Damon 1990 on Trobriand; Meggitt 1974, 185–86, 191; Wiessner and Tumu 1998, 22 on the Mai Enga). The objects used as gifts are either produced by subordinate local group members (pigs by Enga or Maring women or *tugudu* cloth by young Tiv men) or they are imported through the local chiefs' personal relations with long-distance traders (Douglas and Isherwood 1996; Guyer 2004 on the Tiv).

Politically ambitious men in a local group try to outdo each other in gift exchange with the aim of acquiring a higher status. This competition rarely takes the form of a direct confrontation of rivals, as in *potlatch* among the Kwakiutl (Boas 1921). More often the competition is indirect in the form of exchanging more gifts and acquiring more exchange partners in other local groups than rivals, as among Trobriand chiefs (Singh Uberoi 1971), or through the rivals' differing contributions to their village's alliance feasts, as among the Mai Enga (Meggitt 1974; Wiessner and Tumu 1998). The more uneven the quantity of gifts donated, the greater the status difference between the rivals.

Among the Mai Enga, every would-be big-man in a village tries to find as many *te*-partners as possible, because his power and status within the group depends on it. The pigs he has at his disposal come from his own polygamous household, with each wife able to care for about ten pigs. He also receives pigs from members of his local subclan or clan, usually from younger men who hope to be supported by the big-man in finding a wife in turn. He will also receive pigs from 'rubbish men', i.e. men without any relatives in the village. In addition, he lends pigs for rearing to households in villages that

12 Various forms of compensation may be distinguished: a) compensation of allies for casualties in a common war campaign (where one is not the perpetrator, but still responsible), b) compensation of relatives of a victim killed by the perpetrator, if the victim comes from another village, c) compensation of relatives of a victim in one's own village, where one is under the authority of the older men.

are not involved in *te*-exchange. They receive a few piglets as payment for their work. Most of the pigs, however, he receives from his matrilateral relatives and their matrilateral relatives, so that he has as many pigs as possible at his disposal in the village at the time of a feast. The big-man also tries, with promises and threats, to convince the other members of his patriclan to reduce their gifts to affinal relatives in local groups with which his own group is at enmity and to give their pigs to marriage relatives in allied groups instead. The big-man of a patriclan has to achieve his goals in the context of a rivalry with smaller big-men, the leaders of the subclans of his patriclan, who may have diverging preferences in the local group's alliance policy. The higher the difference to the contributions of rivals to alliance feasts, the greater the status difference will be. The intra-group competition for offices, prestige, and status, thus, runs through the provision of larger quantities of alliance goods than rivals' (Meggitt 1974; Wiessner and Tumu 1998).

The position of a chief — like that of local gerontocrats in tribal societies without a paramount chief — is constituted and legitimized by kinship and marriage relations. A Kachin chief owes his position to the fact that he belongs to the oldest branch of the first kinship group who came into the region. As he is the representative of the oldest lineage (or subclan) in the region, he must perform rituals for the local spirits, a task for which he receives goods from all the villagers of his domain. A Kachin chief further increases the quantity of gifts (in the form of bride wealth) by strategically marrying off the women of his subclan into all the villages of his domain and thus bringing the villages of his sons-in-law into dependency. Furthermore, a chief leads war campaigns during which he may capture war slaves and permanently control long-distance trade routes and thus will have more wealth to distribute at merit feasts and recruit more followers than his rivals (Friedman 1998; Leach 1954).

Conversely, on Trobriand Island, the chief of the dominant matri-subclan marries women from all the other villages in his domain and benefits from the tributes of his numerous brothers-in-law whose subclans supply *urigubu* yams to their sisters — and indirectly to their husband, the polygamous chief. Furthermore, the chief receives goods from his matrilateral nephews who want to succeed him. With all these goods, the chief will equip *kula* expeditions: pay craftsmen to produce ships and *kitoum* which then are converted into gifts (*vaygu'a*) as well as finance wars and pay sorcerers. In the *kula* exchange itself, a chief may receive *pokala* (pigs, shells, axes) from his *kula*-partners who desperately want a particularly valuable *vaygu'a* (Malinowski 1922; Powell 1960; 1969a; 1969b; Singh Uberoi 1971).

Three Dimensions of Gift Exchange

To clarify my position, it makes sense to distinguish three different dimensions of gift exchange (Befu 1977, 259–60): a) cultural conceptions and classifications, b) the 'rules of the game', and c) political interests of exchange actors.

Cultural conceptions and classifications of objects, persons, and situations undoubtedly play a role in gift exchange (Strathern 1988; 1992). However, these cultural concepts and models cannot explain the complexity and the various contexts of gift transactions. And they cannot explain why actors reciprocate gifts in the first place, as Mauss (1923–1924) has noticed.

Secondly, there are the 'rules of the game', norms of gift exchange that cannot be reduced to interests: for playing the 'gift game', you should donate a gift, your counterpart should accept and (with a delay) reciprocate it. There are other rules or conventions for how to play the game, under what circumstances (mostly ceremonial), what kind of gifts and in what quantities you should give, and concerning the time schedule of the gift transactions.

In addition to these 'rules of the game', there is, thirdly, the strategic calculating of the exchange actors, the 'playing of the game' (cf. Blau 1964; Bourdieu 1994; 1997).[13] The interests involved in gift exchange are not economic interests but genuine political interests: it is about getting more political support, acquiring more military allies, and higher status positions, i.e. about getting more (relational) power than rivals. These political interests of transactors in gift exchange are absolutely crucial. The donor decides whether he wants to start an exchange of gifts and with whom. The recipient decides whether or not to accept a gift and when and whether to return it (or pass it on). Gifts are not always accepted and not always reciprocated, although actors do not dispute the validity of the rules of the game. And: Why should somebody reciprocate gifts when no supra-local sanctioning instance can enforce compliance with this norm in the first place? Moreover, it is not uncommon for conflicts to arise in the context of gift exchange, since all participants are looking for their own advantage.

This leads us to the last alleged 'enigma of the gift' (Godelier 1996).

13 Berthoud and Sabelli (1976), Caillé (1994; 2000; 2018) and others see gift exchange as largely governed not by interest, but by 'irrational motives' or 'altruism' (see critique in Helbling 2021, 218–28).

'Voluntary and Obligatory' Aspects in Gift and Commodity Exchange

Some theorists are stuck on Mauss's statement that gift exchange is 'at the same time voluntary and obligatory'. According to Testart (2001; 2013; Darmangeat 2016), reciprocity in gift exchange — in contrast to commodity exchange — is not legally enforceable (hence: voluntary). It is true, that a bookseller may call the police, if I have not paid for a book, but Uncle Fritz cannot sue me for not having reciprocated his Christmas present. However, what does the bookseller do if there are no police as is the case in non-state societies? Even in commodity exchange (the supposed realm of legal obligation to pay for goods received), there is always the alternative: either to pay for a good or to take it without paying (Hirshleifer 1978) — see also 'negative reciprocity' (Sahlins 1965). The 'legal obligation to pay for goods received' remains wishful thinking in societies without a state where judicial rules can only be enforced within the villages, but not between villages. That is why commodity exchange is almost impossible without gift exchange and relations of kinship, marriage, or friendship between the trading partners. On Trobriand, *vava*-trade only takes place if relations are good; if not, *wasi*-gifts have to be exchanged, in order to restore peace and to resume *vava*-trade. However, if one of the trading partners does not pay, their relationship is terminated and may easily turn into enmity. The refusal to pay for a commodity leads to the same result as the refusal to reciprocate a gift: the termination of a transactional relationship, i.e. 'private or public war' (Mauss 1923–1924).

What about the 'voluntary and obligatory' aspects in gift exchange? Gift exchange can be conceived as a game that one is interested to play with somebody or not. This is the voluntary part, highlighted by Testart. But in case you want to play the game because you are interested in a relationship with the exchange partner, you have to abide by the rules of the game. That is the obligatory part. Exchange partners will play the game if and as long as both are interested in the mutual relationship that is established, strengthened, or repaired through gift exchange. If one of the exchange partners is no longer interested in the relationship, the game of exchanging gifts is no longer played, at the cost of emerging mistrust and hostility. 'Enfin, ces prestations et contre-prestations s'engagent sous une forme plutôt volontaire, par des présents, des cadeaux, bien qu'elles soient au fond rigoureusement obligatoires, à peine de guerre privée ou publique' (Mauss 1923–1924, 9).

The obligation to reciprocate a gift is thus being observed, not because of the punishment by a sanctioning higher authority (a state) which does not exist, or because of the fear of a spiritual power such as the *hao*, but because not reciprocating would terminate a relationship (of marriage and/or alliance) in which one is interested. It is the political interest of exchange partners, stupid!

Works Cited

Bailey, F. 1969. *Stratagems and Spoils* (London: Blackwell).

Barth, F. 1967. 'Economic Spheres in Darfur', in R. Firth (ed.), *Themes in Economic Anthropology* (London: Tavistock), pp. 149–74.

Befu, H. 1977. 'Social Exchange', *Annual Review of Anthropology*, 6: 225–81.

Berthoud, G., and F. Sabelli. 1976. *L'ambivalence de la production: logiques communautaires et logique capitaliste* (Geneva: Institut d'études du développement).

Blau, P. 1964. *Exchange and Power in Social Life* (New York: Wiley).

Bloch, M., and J. Parry. 1989. 'Introduction', in J. Parry and M. Bloch (eds), *Money and the Morality of Exchange* (Cambridge: Cambridge University Press), pp. 1–32.

Boas, F. 1921. *Ethnology of the Kwakiutl* (Washington, DC: Bureau of American Ethnology).

Bohannan, P. 1955. 'Some Principles of Exchange and Investment among the Tiv', *American Anthropologist*, 57: 60–70.

———. 1959. 'The Impact of Money on an African Subsistence Economy', *Journal of Economic History*, 19: 491–503.

Bohannan, P., and G. Dalton. 1962. 'Introduction', in P. Bohannan and G. Dalton (eds), *Markets in Africa* (Evanston: Northwestern University Press), pp. 1–16.

Bourdieu, P. 1994. *Raisons pratiques: sur la théorie de l'action* (Paris: Seuil).

———. 1997. *Méditations pascaliennes* (Paris: Seuil).

Caillé, A. 1994. *Don, intérêt et désintéressement* (Paris: La découverte).

———. 2000. *Anthropologie du don: le tiers paradigme* (Paris: Desclée de Brouwer).
———. 2018. 'Qu'est-ce qui ne va pas avec le don chez Bourdieu? Le don n'est pas un acte économique mais un opérateur politique', *Revue du MAUSS*, 52: 74–88.
Camerer, C. 1988. 'Gifts as Economic Signals and Social Symbols', *American Journal of Sociology*, 94: 180–214.
Chapman, A. 1980. 'Barter as a Universal Mode of Exchange', *L'Homme*, 3: 33–83.
Dalton, G. 1965. 'Primitive Money', *American Anthropologist*, 61: 44–65.
Damon, F. 1990. *From Muyuw to the Trobriands: Transformations along the Northern Side of the Kula Ring* (Tucson: University of Arizona Press).
Darmangeat, C. 2016. 'Don, échange et autres transferts: Formes simples, hybrides et composée', *L'Homme*, 217: 21–43.
Douglas, M., and B. Isherwood. 1996. *The World of Goods* (London: Routledge).
Evans-Pritchard, E. 1940. *The Nuer: A Description of the Modes of Livelihood and Political Institutions of a Nilotic People* (Oxford: Clarendon).
Firth, R. 1959. *Economics of the New Zealand Maori*, rev. edn (Wellington: Government Printer).
Fortune, R. 1932. *Sorcerers of Dobu: The Social Anthropology of the Dobu Islanders of the Western Pacific* (London: Routledge).
Freeman, D. 1970. *Report on the Iban* (London: London School of Economics Press).
Friedman, J. 1998. *System, Structure, and Contradiction* (Walnut Creek: AltaMira).
Gell, A. 1992. 'Inter-tribal Commodity Barter and Reproductive Gift Exchange in Old Melanesia', in C. Humphrey and S. Hugh-Jones (eds), *Barter, Exchange and Value* (Cambridge: Cambridge University Press), pp. 142–68.
Gewertz, D. 1978. 'Tit for Tat: Barter Markets in the Middle Sepik', *Anthropological Quarterly*, 51: 37–44.
———. 1983. *Sepik River Societies: A Historical Ethnography of the Chambri and their Neighbours* (New Haven: Yale University Press).
Gintis, H., E. A. Smith, and S. Bowles. 2001. 'Costly Signalling and Cooperation', *Journal of Theoretical Biology*, 213: 103–19.
Godelier, M. 1977. 'Salt Money and the Circulation of Commodities among the Baruya of New Guinea', in J. Kahn (ed.), *Perspectives in Marxist Anthropology* (Cambridge: Cambridge University Press), pp. 127–51.
———. 1996. *L'énigme du don* (Paris: Fayard).
Görlich, J. 1992. 'Gabentausch und Tauschhandel in Melanesien als strategische Interaktion', *Sociologus*, 42: 24–42.
———. 1996. 'Ceremonial Gift Exchange and Barter in Melanesia: A Game Theoretical Differentiation', *Angewandte Sozialforschung*, 20: 37–46.
Graeber, D. 2001. *Toward an Anthropological Theory of Value: The False Coin of our Own Dreams* (New York: Palgrave).
———. 2011. *Debt (Updated and Expanded): The First 5,000 Years* (London: Melville House).
Gregory, C. 1982. *Gifts and Commodities* (London: Academic Press).
———. 1994. 'Exchange and Reciprocity', in T. Ingold (ed.), *Companion Encyclopedia of Anthropology* (London: Routledge), pp. 911–40.
———. 1997. *Savage Money: The Anthropology and Politics of Commodity Exchange* (Amsterdam: Harwood).
Guyer, J. 2004. *Marginal Gains: Monetary Transactions in Atlantic Africa* (Chicago: University of Chicago Press).
Haas, J. 1990. 'Warfare and the Evolution of Tribal Polities in the Prehistoric Southwest', in J. Haas (ed.), *The Anthropology of War* (Cambridge: Cambridge University Press), pp. 171–89.
Hart, K. 1986. 'Heads or Tails? Two Sides of the Coin', *Man*, 21: 637–56.
———. 2005. 'Money: One Anthropologist's View', in J. Carrier (ed.), *A Handbook of Economic Anthropology* (Cheltenham: Elgar), pp. 160–93.
Heady, P. 2005. 'Barter', in J. Carrier (ed.), *A Handbook of Economic Anthropology* (Cheltenham: Elgar), pp. 262–74.
Healey, C. 1984. 'Trade and Sociability: Balanced Reciprocity as Generosity in the New Guinea Highlands', *American Ethnologist*, 11: 42–60.
———. 1985. 'New Guinea Inland Trade: Transformation and Resilience in the Context of Capitalist Penetration', *Mankind*, 15: 127–43.
Helbling, J. 2006. *Tribale Kriege* (Frankfurt: Campus).
———. 2021. *Neue Politische Ökonomie einfacher Gesellschaften: Wirtschafts- und politikethnologische Erkundungen* (Wiesbaden: Springer).
Hirshleifer, J. 1978. 'Competition, Cooperation, and Conflict in Economics and Biology', *American Economic Review*, 68: 238–43.
Keeley, L. 1996. *War before Civilization* (Oxford: Oxford University Press).
Kopytoff, I. 1986. 'The Cultural Biography of Things: Commoditization in Process', in A. Appadurai (ed.), *The Social Life of Things: Commodities in Cultural Perspective* (Cambridge: Cambridge University Press), pp. 64–91.
Leach, E. 1954. *Political Systems of Highland Burma: A Study of Kachin Social Structure* (London: Bell).
Malinowski, B. 1921. 'The Primitive Economics of the Trobriand Islanders', *Economic Journal*, 31: 1–15.

———. 1922. *Argonauts of the Western Pacific* (London: Routledge).
———. 1935. *Coral Gardens and their Magic* (London: Allen & Unwin).
Mauss, M. 1923–1924. *Essai sur le don: forme et raison de l'échange dans les sociétés primitives* (Frankfurt: Suhrkamp).
McCarthy, F. 1938–1939. 'Trade in Aboriginal Australia, and Trade Relationships with Torres Strait, New Guinea and Malaya', *Oceania*, 9: 405–58.
———. 1939–1940. 'Trade in Aboriginal Australia, and Trade Relationships with Torres Strait, New Guinea and Malaya', *Oceania*, 10: 80–104, 171–95.
Meggitt, M. 1974. '"Pigs Are our Hearts": The *Te* Exchange Cycle among the Mae Enga of New Guinea', *Oceania*, 44: 165–203.
Modjeska, N. 1985. 'Exchange Value and Melanesian Trade Reconsidered', *Mankind*, 15: 145–62.
Polanyi, K. 1957. 'The Economy as an Instituted Process', in K. Polanyi, C. Arensberg, and H. Pearson (eds), *Trade and Market in the Early Empires* (Chicago: Regnery), pp. 243–70.
Pospisil, L. 1994. '"I Am Sorry I Cannot Kill You Anymore": War and Peace among the Kapauku', in S. Reyna and R. Downs (eds), *Studying Warfare: Anthropological Perspectives* (Amsterdam: Gordon and Breach), pp. 113–26.
Powell, H. 1960. 'Competitive Leadership in Trobriand Political Organisation', *Journal of the Royal Anthropological Institute*, 90: 118–45.
———. 1969a. 'Genealogy, Residence and Kinship in Kiriwina', *Man*, 4: 177–202.
———. 1969b. 'Territory, Hierarchy and Kinship in Kiriwina', *Man*, 4: 580–604.
Rospabé, P. 1993. 'Don archaïque et monnaie sauvage', in *Ce que donner veut dire: don et intérêt. Revue du MAUSS*, 1: 33–59.
———. 1995. *La dette de vie: aux origines de la monnaie* (Paris: La Découverte).
Sahlins, M. 1965. 'On the Sociology of Primitive Exchange', in M. Sahlins (ed.), *Stone Age Economics* (London: Tavistock), pp. 185–275.
———. 1968. *Tribesmen* (Englewood Cliffs: Prentice-Hall).
———. 1974. 'Exchange Values and the Diplomacy of Primitive Trade', in M. Sahlins (ed.), *Stone Age Economics* (London: Tavistock), pp. 277–314.
Schneider, H. 1974. *Economic Man* (New York: Free Press).
Sillitoe, P. 1978. 'Ceremonial Exchange and Trade: Two Contexts in which Goods Change Hands in the Highlands of Papua New Guinea', *Mankind*, 11: 265–75.
Singh Uberoi, J. 1971. *The Politics of the Kula Ring* (Manchester: Manchester University Press).
Smith, A. [1776] 1999. *An Inquiry into the Nature and Causes of the Wealth of Nations* (London: Penguin).
Smith, E. A., and R. Bliege Bird. 2005. 'Costly Signaling and Cooperative Behavior', in H. Gintis, S. Bowles, R. Boyd, and E. Fehr (eds), *Moral Sentiments and Material Interests: On the Foundations of Cooperation in Economic Life* (Cambridge, MA: MIT Press), pp. 115–48.
Strathern, A. 1971. *The Rope of Moka: Big-Men and Ceremonial Exchange in Mount Hagen* (Cambridge: Cambridge University Press).
———. 1988. *The Gender of the Gift* (Cambridge: Cambridge University Press).
———. 1992. 'Qualified Value: The Perspective of Gift Exchange', in C. Humphrey and S. Hugh-Jones (eds), *Barter, Exchange and Value* (Cambridge: Cambridge University Press), pp. 169–91.
Strathern, A., and P. Stewart. 2005. 'Ceremonial Exchange', in J. Carrier (ed.), *A Handbook of Economic Anthropology* (Cheltenham: Elgar), pp. 230–45.
Testart, A. 2001. 'Échange marchand, échange non marchand', *Revue française de sociologie*, 42: 719–48.
———. 2013. 'What Is a Gift?', *HAU: Journal of Ethnographic Theory*, 3: 249–61.
Weiner, A. 1985. 'Inalienable Wealth', *American Ethnologist*, 12: 210–27.
———. 1992. *Inalienable Possessions: The Paradox of Keeping-While-Giving* (Berkeley: University of California Press).
Wiessner, P., and A. Tumu. 1998. *Historical Vines* (Washington, DC: Smithsonian Institution Press).
Znoj, H. 1995. *Tausch und Geld in Zentralsumatra* (Berlin: Reimer).

SABRINA KUTSCHER

6. The Early Bronze Age Rib Ingot Hoard from Oberding (Upper Bavaria)

ABSTRACT Thanks to an excavation prior to construction and subsequent recovery in two blocks of earth (en bloc), the rib ingot hoard of Oberding in Upper Bavaria, discovered in 2014 on the eastern edge of the Erdinger Moos on the fertile loess terrace, can be counted among the best-documented Early Bronze Age hoards. Non-invasive analyses using computer tomography have precisely documented the exact arrangement of the 794 copper ingots in the feature. The ingots were tied into handy bundles of ten and placed in eight larger groups in a separate niche at the edge of an already partially filled pit. In the direct surroundings of the deposit, further pit features and their find material provide an insight into a settlement, which also dates to the late Early Bronze Age. Extensive preliminary work, especially an elaborate restoration and first metallurgical analyses, have already produced important insights that underline the outstanding importance of this find. The copper of the rib ingots comes from different deposits, indicating the significance of the hoard for the interface of the Early to Middle Bronze Age and the changing patterns of commodity transfer and cultural contacts. In order to answer important questions regarding the copper types used, the transfer of goods, value systems, reasons for the deposition, as well as its place of settlement, the rib ingot hoard and its direct surroundings were part of an interdisciplinary research project soon after the excavation. This article gives an overview of the results of the research so far, which could be presented in the form of a poster at the workshop on 'Number and Measure in Prehistory – The Materiality of Mathematical Thinking' at the Institute of Archaeology, University of Zurich (Switzerland) in early May 2020.

KEYWORDS Early Bronze Age, Upper Bavaria, South Germany, hoard, rib ingots, block excavation, computer tomography, number and weight, metallurgical analyses

Introduction

The rib ingot hoard from Oberding belongs to one of the most important find categories of the prehistoric research within the Early Bronze Age: the hoard finds (von Brunn 1995; David 2015, 87–112; Innerhofer 1997, 53–59; Krenn-Leeb 2010, 281–315). Pure hoards made of raw copper in the form of so-called ingots significantly characterize the Early Bronze Age hoard landscape in southern Bavaria and represent the main part of this find category (Möslein 1998–1999, 69–77; 2008, 109–30; 2015, 113–22). They can be found in moors, lakes, rivers, and rock crevices, but also — as in the case of Oberding — in the middle of the contemporary settlement landscape on fertile mineral soils (David 2015, 87–112).

The archaeological hoard landscape of southern Bavaria has been one of the most well-researched

Sabrina Kutscher University of Munich, LMU. From 2010 to 2016 studies of pre- and early historical archaeology at the LMU Munich and FAU Erlangen, with a focus on the Neolithic and Bronze Age as well as archaeobotany. Since 2017 doctoral studies at the LMU Munich with the topic 'The Early Bronze Age Settlement Area in the Erding Land. An Archaeological and Scientific Examination of Selected Sites' under the supervision of Carola Metzner-Nebelsick. 2019 to 2020 archaeological assistant in the DFG project 'The Early Bronze Age Rib Ingot Hoard of Oberding, District of Erding and its Surroundings. Archaeological and Scientific Evaluation' under the direction of Carola Metzner-Nebelsick.

Number and Measure in Prehistory: Gift, Exchange, and the Transfer of Goods, ed. by Philippe Della Casa and Aleksander Dzbyński, WSA, 6 (Turnhout, 2024), pp. 79–91.
BREPOLS PUBLISHERS
DOI 10.1484/M.WSA-EB.5.137131

in central Europe since the end of the 1970s and the beginning of the 1980s due to of Frauke Stein (1976; 1979), Manfred Menke (1978–1979), and Karl-Friedrich Rittershofer (1983) (see also Neumann 2015).[1] Since these editions of hoard finds, the corpus of Early Bronze Age hoards in Bavaria has increased notably (David 2015, 87–112).[2] The majority of the old finds are chance finds, which were discovered during earthworks, e.g. in gravel pits and peat ditches or in the course of ploughing. Particularly in the case of the old finds, documentation of the circumstances of the features and their description is usually missing, which impairs the recontextualization of these finds to this day (Möslein 2008, 109–30). The recontextualization of not or insufficiently documented hoards has therefore been an important contribution to hoard research for several years, as Daniel Neumann, among others, has vividly demonstrated in his work on ritual landscapes (Ballmer 2010, 120–31; Hansen and others 2012; Neumann 2015). Stephan Möslein (2015, 113–18), for example, was able to show for the total frequency of Bronze Age finds in south-eastern Upper Bavaria that most of the finds up to the 1970s were chance finds, but since the 1980s, a strong increase in hoards could be documented by specific searches with metal detectors, with a decrease in chance finds. Only a small number of hoards, which were discovered by detectorists, such as the hoard of rib ingots of Adelpolt in the district of Ebersberg in 2019, are left *in situ*, reported to the Bavarian State Office for the Preservation of Monuments, and scientifically documented and recovered in the course of an archaeological excavation (Möslein 2015, 116–18). Hoard finds that were discovered and professionally documented during regular excavations in the last few years, usually prior to construction, such as the hoard of Greding, Roth district, discovered in 2002, are still an even greater exception (Nadler 2003, 36–38). The discovery of the so far largest hoard of rib ingots of the Early Bronze Age in Oberding in the course of a regular excavation is therefore an absolute stroke of luck for hoard research.

Within the framework of the workshop on 'Number and Measure in Prehistory – The Materiality of Mathematical Thinking' under the direction of Aleksander Dzbyński and Philippe Della Casa at the Institute of Archaeology, Dept. of Prehistoric Archaeology, University of Zurich (Switzerland), an overview of the previous results of the investigation could be given in the form of a poster. The following article presents the written version of this poster, which focuses on the presentation of the feature and the composition of its finds.

The Discovery and Research of the Oberding Hoard

The village of Oberding is located on the western flank of a loess-covered, crack-Ice-Age high terrace about six kilometres north-west of the city of Erding in Upper Bavaria. The almost flat high terrace rises like an elongated peninsula tongue-like to the north into the low moor landscape of the Erdinger Moos and is characterized by fertile parabrown soils. It is therefore not surprising that this loess terrace has the highest density of soil monuments in the entire district of Erding and that the first remains of settlement can already be attributed to the Linear Band Culture (5500–4900 BC) (Kopecky-Hermanns and Krause 2013, 14–15).

The site Oberding 'Sonnenstraße' is located in a ground monument that has been designated as a 'settlement of prehistoric and early historical times' since the mid-1950s.[3] Although these areas are recorded in the list of known or probable sites by the Bavarian State Office for Monument Protection and Cultural Heritage (BLfD), no archaeological investigations have been carried out in the immediate surroundings since the 1970s and 1980s despite continuous settlement expansion (Stadt Erding 2017, 43). In this respect, it was fortunate that in spring 2014 one of the last remaining gaps in the 'Sonnenstraße' could be archaeologically examined prior to construction works, in the course of which the Early Bronze Age hoard was discovered. In order to ensure an optimal documentation of the complex feature situation, the hoard was blocklifted in two blocks and brought to the Bavarian State Office for Monument Protection and Cultural Heritage (BLfD) in Munich for further investigations (Gasteiger 2017,

1 Early Bronze Age ingot hoards are not limited to the northern Alpine foothills, as the studies of Alexandra Krenn-Leeb (2008, 163–196; 2010, 281–315) and Wolfgang Neugebauer (1977), among others, have shown for Austria. For Bohemia, Early Bronze Age hoards including the ingot depots have been presented by Václav Moucha (2005) and recently by Ondřej Chvojka, Luboš Jiráň, and Milan Metlička (until 2013) (Chvojka and others 2017a; 2017b).

2 Stephan Möslein in particular (1997, 55–58; 1998a, 73–74; 1998–1999, 69–77; 2008, 109–30; 2015, 113–22) has been working on Bronze Age hoard finds in the eastern Bavarian Alpine foreland for almost twenty years, with a special focus on ring and rib ingot hoards.

3 Field surveys in the 1950s and 1960s by Siegfried Ried, a teacher from Oberding, which yielded Early Bronze Age pottery finds, as well as positive soil characteristics of features on aerial photographs have since indicated an Early Bronze Age activity zone (Press 1953; Ried 1968, 178).

60–61; Krause and others 2018a, 167–69; Kutscher 2017, 423–34; Leicht and Stöckl 2015, 39–42).

The hoard of Oberding will be comprehensively analysed within the framework of a cooperation project between the Bavarian State Office for Monument Protection and Cultural Heritage (BLfD) in Munich, the city of Erding (Museum Erding), as well as the Institute of Pre- and Proto-historic Archaeology and the Archaeology of the Roman Provinces at the Ludwig-Maximilians-University Munich (May 2015–July 2017) and a subsequent DFG project (February 2019–April 2020) under the direction of Carola Metzner-Nebelsick at LMU Munich.[4]

From May 2015 to July 2017, the finds and block excavations were restored, uncovered, and archaeologically documented in the BLfD's restoration workshops (Gasteiger 2017, 60–61). We decided to leave 173 ingots *in situ* in the block of earth and thus make them available for future research. Parallel to the restoration work and the archaeological recording of finds, extensive scientific investigations were carried out, including geoarchaeology, archaeobotany, archaeozoology, lipid analyses, XRF and lead isotope analyses, and ^{14}C dating. Within the framework of a master's thesis under the supervision of Carola Metzner-Nebelsick, the first archaeological results were presented by the author in July 2016 and published in excerpts (Krause and others 2018b, 22–23; Kutscher 2017, 423–34). Already in July 2017, the entire find complex could be permanently integrated into the archaeological department of the Museum Erding. At the same time, the current state of work and research of the individual project participants could be presented to experts and the general public at a conference at the Museum Erding and made accessible to a general public in a popular science exhibition volume of the Museum Erding publication series (Later 2017, 83–85; Stadt Erding 2017).

So far, only the hoard itself and the pit in which it was deposited have been investigated. The DFG project therefore focuses on the archaeological examination of the directly adjacent feature context, i.e. the entire excavation of the Oberding 'Sonnenstraße' find site and the southern connecting area. These form the immediate surroundings of the hoard.[5] Since the features continued beyond the southern excavation boundary of the Oberding 'Sonnenstraße' site, the area adjoining the site to the south was archaeologically investigated in late summer 2017 in a further construction project. The features and finds of this area and the area excavated in 2014 indicate a settlement of the late Early Bronze Age in the immediate surroundings of the hoard find.

As a complete, closed find, the rib ingot hoard from Oberding offers great potential for answering a multitude of questions. Both cultural-historical and economic-historical aspects will be addressed. In addition, the environmental settings, both nearby and further away from the site, will be investigated to gain a better picture of the Early Bronze Age cultural landscape. Thus, the interweaving of artefacts and features with the living world and society of the Early Bronze Age will be shown. The assemblage of objects with similar weight suggests its function as distinctive weight values (units) within a pre-monetary system of exchange, and in particular in the spheres of divine as well as social interaction. An important aspect of the investigation, therefore, is the examination of the weight values as well as the material quantity and composition of the ingots from the hoard of Oberding.

4 Institute of Pre- and Proto-historic Archaeology and the Archaeology of the Roman Provinces at the Ludwig-Maximilians-University Munich, 2019, Institute of Pre- and Proto-historic Archaeology and the Archaeology of the Roman Provinces, Munich, Project Oberding <https://www.vfp-archaeologie.uni-muenchen.de/forschung/vorfrueh/projekt_oberding/index.html> [accessed 9 May 2022].
Archaeology (LMU Munich): Prof. Dr Carola Metzner-Nebelsick, Sabrina Kutscher MA; Bavarian State Office for Preservation of Monuments in Munich (BLfD): Prof. Dr C. Sebastian Sommer + Dipl.-Rest. Stephanie Gasteiger; Museum Erding: Harald Krause M.A., Dipl.-Ing. Wilhelm Wagner; Excavation: Archäologisches Büro Anzenberger & Leicht (excavation 2014), X-Cavate Archaeology (excavation 2017); Restoration (BLfD): Dipl.-Rest. Jörg Stolz, Dipl.-Rest. Thomas Stöckl; Archaeobotany: Dipl.-Biol. Barbara Zach (Bernbeuern), Sabrina Kutscher MA; Archaeozoology (LMU Munich): Prof. Dr Wolf-Rüdiger Teegen, Silvia Eccher MA; Geoarchaeology: Dipl.- Geogr. Britta Kopecky-Hermanns (Aystetten), Dr Dagmar Fritzsch (Goethe-University Frankfurt), Dr Holger Rittweger (Waldbrunn), Dr Werner Häusler (TUM Weihenstephan), Palynological investigations Dr Philipp Stojakowits (University of Augsburg), PD Dr Michael Peters (LMU Munich); Dendroarchaeology (BLfD): Franz Herzig; Metal analysis: Prof. Dr Ernst Pernicka (CEZA Mannheim), Dipl.-Ing. Björn Seewald (BLfD); ^{14}C dating: Dr Ronny Friedrich (CEZA Mannheim); lipid analysis (Eberhard Karls University Tübingen): Dr Jun.-Prof. Cynthianne Spiteri; 3D computed tomography (Fraunhofer Institute for Integrated Circuits IIS Fürth): Nils Reims; 3D printing of the small block recovery / structure from motion for small and large block recovery: Formwerk3D Hannover, Archäologische Illustrationen.

5 In order to be able to better classify the site of the rib ingot hoard into its regional surroundings, the author is researching in the context of her doctoral thesis, following on from the DFG project, the Early Bronze Age settlement landscape between the Isar and Inn Rivers in the region of the Erdinger Land, starting from the loess terrace of Oberding. Institute of Pre- and Proto-historic Archaeology and the Archaeology of the Roman Provinces at the Ludwig-Maximilians-University Munich, 2019, Institute of Pre- and Proto-historic Archaeology and the Archaeology of the Roman Provinces, Munich, Sabrina Kutscher MA <https://www.vfp-archaeologie.uni-muenchen.de/personen/wiss_ma/sabrina_kutscher/index.html> [accessed 9 May 2022].

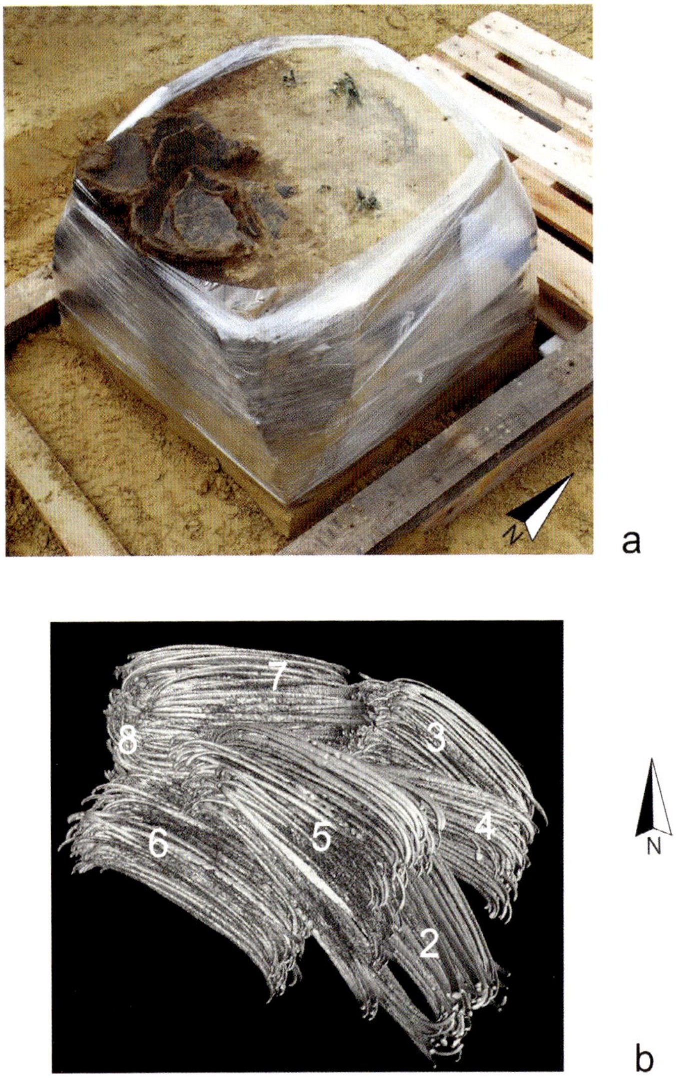

Figure 6.1. The large block to be recovered was exposed all around in a rectangular shape with 90 cm edge length and 60 cm height (1a). The situation of the hoard was documented by 3D CT images before the exposure (1b), whereby the position and shape of the rib ingots in bundles and groups were already clearly visible (BLfD, Thomas Stöckl 2014, Fraunhofer EZRT 2016).

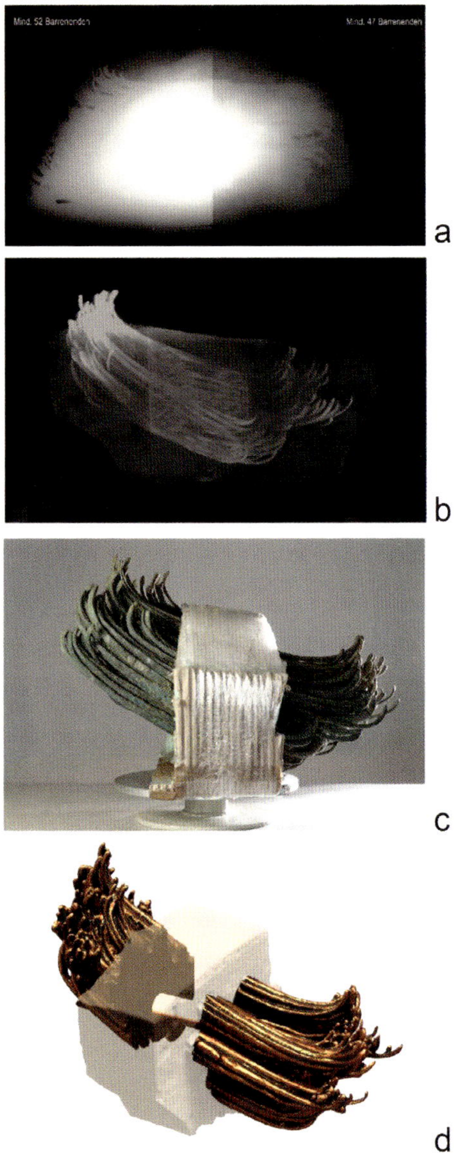

Figure 6.2. Based on an X-ray image, the extent of the small block was estimated to be at least fifty-two rib ingots (2a). The CT images taken at the Fraunhofer EZRT in Fürth made it possible to visualize the actual number of 141 rib ingots (2b). During the exposure of the ingots, 250 photos of the small block excavation were taken, a 3D model was calculated from these photos, and a 3D print for the museum presentation was produced (2c–d) (BLfD, Thomas Stöckl 2014; Fraunhofer EZRT, Nils Reims 2015; BLfD, Jörg Stolz 2016; Eik Jagemann 2017).

Non-invasive Examination Methods

The majority of the Oberding rib ingots had been deposited in a cave-like niche (feat. 1b), which bordered a pit (feat. 1a) at the north-eastern edge. Only the highest group of ingots (group 1) was located at the edge of pit 1a (feat. 1a). Those ingots, which lay about 30 cm higher than the remaining ingots, were recovered separately as a small block. The edge of pit 1a revealed a kind of broken sherd pavement and the adjacent remaining ingots (groups 2–8). They were lifted in a larger block excavation (90 × 90 × 60 cm) (Fig. 6.1a). To clarify their stratigraphic relation, the restoration work was carried out under geoarchaeological supervision.

In order to obtain a precise overview of the structure and arrangement of the artefacts in the compactly laid down ensembles even before their uncovering, non-invasive imaging techniques were used (Fig. 6.1b) (Haßmann and others 2014, 743–52).

A first insight was provided by an X-ray examination of the small block excavation (group 1). However, due to the high density and the mutual superposition of the finds, the analysis only gave a diffuse image (Fig. 6.2a). Due to the high density of the large block excavation (approx. 800 kg), the use of a normal X-ray machine at the BLfD was not possible. At the Fraunhofer Institute in Fürth, a CT image of the small block excavation was taken (Fig. 6.2b). The result was impressive: the three-dimensional image already allowed the shape and position of individual ingots and their exact number to be determined. During the large block recovery, the XXL-CT system was at the limit of its capacity. For this reason, the area around the shattered large vessel had to be documented first (including Structure-from-Motion (SfM) images), and was then removed accompanied by archaeological documentation. Only after the reduction of the block size, was a three-dimensional image of the large block excavation successfully created. Excellent results were obtained, which served as a working basis and orientation aid in the further course of the excavation work in the restoration lab. The CT already allowed a fairly accurate assessment of the position of ingots within the block (Stolz 2016, 51–53). For example, the rib ingots had been deposited 'very organized' in several groups (Fig. 6.1b). Both the SfM images of both block excavations and the CT images could be used for the museum presentation in the permanent archaeological exhibition at the Erding Museum (Stadt Erding 2017, 109–11). For example, a realistic haptic model of the uppermost rib ingot group was created from the virtual 3D data as a 3D print. This three-dimensional printout shows the position of each individual object clearly and in detail in its original position (Fig. 6.2c–d).

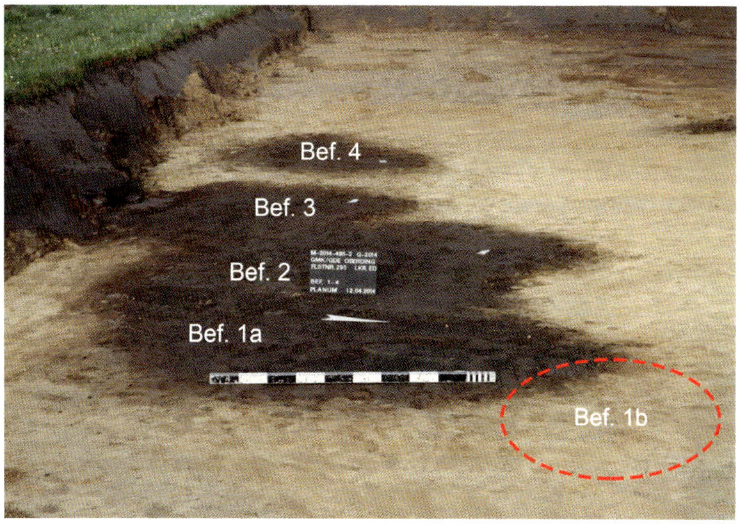

Figure 6.3. Photo of the connected pits 1a to 3 and the adjacent pit 4. The red circle marks the location of the hoard pit (feat. 1b) (Jakob Leicht 2014).

Taphonomy

The geoarchaeological investigations and the CT-documentation can be combined for the description of the find taphonomy (Kopecky-Hermanns and Krause 2018, 41). Starting at the edge of pit 1a, a cave-like niche (feat. 1b) was deepened into the loess soil (Fig. 6.3). At this point in time, according to the stratigraphy, pit 1a was already partially filled in. The niche (80 × 80 × 60 cm) was adapted to the curved pit edge of feat. 1a in this area. The opening of the hoard niche directed to pit 1a measured 60 cm. This was used to fill the pit with the rib ingots. A vertical opening from the earth's surface was not visible.

The rib ingots were tied together in bundles of ten and placed in eight groups, seven of which were placed in the niche (feat. 1b). The lowest groups 3 and 8 lay side by side in the rear area of the hoard niche on the loess. Group 4 was placed above group 3 and group 7 above group 8. In the front area of the feature, groups 2 and 6 were also placed side by side and group 5 on top of these groups. When the fifth group was deposited, there was no more space in the niche for further rib ingots. Groups 2 and 6 were already very close to the edge of the niche and their ends protruded into the edge of feat. 1a. Almost all the groups were covered with a layer of loess, and were deposited cleanly separated from each other. The further one approached the hoard niche opening and thus the adjacent, highly humusy pit 1a during the deposition process, the more inhomogeneous became the sediment surrounding the ingot groups. The niche had obviously been dug especially for the deposition and was then filled in again with the previously removed soil material. The absence of alluvial layers in the hoard niche indicates that the pit could not have been open for a long time (i.e. years).

In direct spatial connection with the hoard, a large vessel was found. It was crushed in the state of discovery, with the opening of the vessel oriented towards the rib ingots. In the course of the restoration it could be almost completely reassembled (Fig. 6.4).

However, this is not an intentionally created sherd pavement. The vessel was most likely originally placed upside down in front of the niche opening. The highest group 1 of ingots may not have been deposited together with the seven other groups, but rather an indefinite time span later. It was located directly in front of the niche in the edge of pit 1a and here partly above the large vessel which had probably already been broken up at this time (Fig. 6.5).

Still, the question of the depth of the depositing process in the Bronze Age remains. Is it one 'hoard' in the narrower sense, i.e. was it created in one act or can we accommodate several small deposition events in a very small space? Given this, there would have been a more or less large time window between the depositing of the individual groups. Based on the feature situation and the pedological-geoarchaeological research results, several depositing scenarios would have resulted in a similar feature picture in the backfill. In this respect, there are limits to answer the question. The most probable scenario for groups 2 to 8 is a one-time deposition process in which the large vessel represents the final step. In another scenario, groups 2 to 8 were deposited one after the other with short interruptions into the niche that had been open for a longer period. This could have been days to several months (depending on season and weather). The large vessel could have served as a closure capsule or marker (Kopecky-Hermanns and Krause 2017, 65–77; 2018, 41). Group 1 of the rib ingots lying in front of the hoard niche and partly on top of the shattered large vessel was probably deposited later, knowing that the niche was already filled to the brim with rib ingots and that it was important to deposit this group in close proximity to the other ingots. Only after group 1 had been deposited, pit 1a, where the hoard niche (feat. 1b) is in direct vicinity to it, was completely filled in. Furthermore, it cannot be excluded that more ingots were originally deposited outside the hoard niche, as indicated by two ploughed out ingot fragments in the colluvium above feat. 3. This means that people were aware of the site of the deposit for a longer period, during the late Early Bronze Age (1750–1650/1600 BC), and that the spot had a special significance (Fig. 6.6).[6]

The Composition of the Hoard

Based on the X-ray images of the small block excavation, it was initially assumed that a hoard of a smaller size (approx. three hundred pieces) was used. Thanks of the CT images; the number was finally estimated as 807 ingots (141 ingots from the small block and 666 ingots from the large block). An exact separation of the ingot ends was not always perfectly possible due to image noise and high find density and thus thirteen ingots too many had been counted during the preliminary investigation. Only after the uncovering, the exact number of 794 objects was determined (Blana and others 2016, 48–50; Stolz 2016, 51–53). Except for the absence of end tips and cracks on a few objects, the ingots are completely preserved. However, two ingot fragments with fresh fractures were found in the colluvium above feat. 3. It is very likely that in the last decades some higher ingots had been ploughed out, so that the original number of ingots was probably larger.

All the ingots can be assigned to the Bermatingen type (1750–1650/1600 BC) (Hájek 1954, 144; Stein 1976, 40–46). On average, the ingots are 30 cm long. The ingot's body is elongated and the ingot ends are evenly curved. In some cases, the ends are folded in at almost right angles. The cross-sections are either flat D-shaped or triangular in various forms. Experimental casting tests by Jörg Stolz (BLfD) with a soapstone mould as well as characteristics on the original have shown that the open hearth casting method was used (Fig. 6.7). However, contrary to the view of earlier research, subsequent bending or forging of the ingot ends, as the sole shaping process is not absolutely necessary. The typical ingot shape can also be produced in other

6 The pottery, which can be used for a relative chronological classification, was mainly concentrated in pit 1a. Some forms can be assigned to the pottery group Sengkofen/Jellenkofen; however, some pieces show contacts to the groups Věteřov-Böheimkirchen and Maďarovce, which are widespread in eastern Austria and the north-western Carpathian region (Bátora 2000; Honig 2008; Möslein 1998b, 37–106; Neugebauer 1977, 31–207; Wullinger 2004, 9–104; Bankus 2004). Both the typochronological dating of the Bermatingen type rib ingot, the raw copper from the Mitterberg region, and ^{14}C-dated organic finds (LaborNr. MAMS 26711: 3425 ± 27,cal BC 1752–1688 (2-sigma). Laboratory No. MAMS 26712: 3359 ± 28, cal BC 1738–1536 (2-sigma). LaborNr. MAMS 26714: 3393 ± 26, cal BC 1747–1623 (2-sigma)) from pit 1a date the entire find complex to the late Early Bronze Age (according to Walter Ruckdeschel A2b/c; according to Stephan Möslein FB III: Massy 2018; Möslein 1998, 37–106; Ruckdeschel 1978).

Figure 6.4. The 33 cm high, almost completely reconstructed large vessel was located directly in front of the opening of the hoard niche, formerly probably upside down. Its form and decoration date it to the end of the Early Bronze Age (1750–1600 BC) (BLfD, Jörg Stolz 2017).

Figure 6.5. Small and large block recovery in the find situation graphically superimposed. The highest bundle group 1 (small block) was located outside the hoard niche (feat. 1b) at the edge of pit 1a partly above the shattered large vessel. The remaining seven groups of bundles lay on top of each other and stacked next to each other in the separate hoard pit (BLfD, Jörg Stolz 2016).

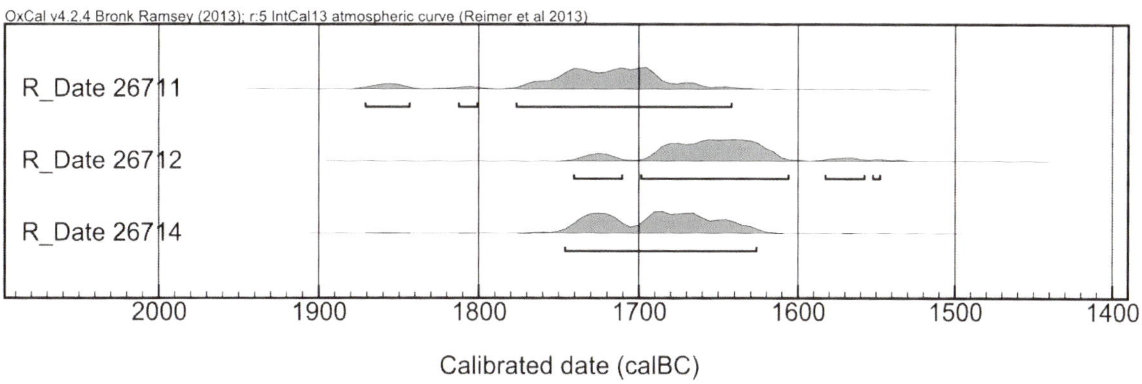

Figure 6.6. Feature 1a: calibrated ^{14}C data of two charred cereal grains (R_Date 26711 and R_Date 26712) from the pit bottom area and one animal bone (R_Date 26714) from the upper filling area (LMU Sabrina Kutscher 2020; data: Ronny Friedrich, CEZA Mannheim 2016).

Figure 6.7. Experimental recasting of a Bermatingen type rib ingot in a movable soapstone mould (Olga Emgrund 2017).

Figure 6.8. In contact with the copper corrosion, organic residues on the ingots, which could be determined as tree bast, have been preserved (8a). Reconstruction of the three proven lacing variants of the bundles of ten (8b) (BLfD, Jörg Stolz 2016 and 2017).

ways.[7] In the case of experimental castings, one variant has proved to be the most likely: the casting mould would be inclined towards the tip of the ingot. The liquid metal is poured into one end of the ingot and then the mould is aligned horizontally. The melt flows to the centre of the ingot. The pouring is repeated on the opposite side (Stolz 2017a, 150–53; 2017b, 14–19). Casting lines that follow the ingot shape and run as grooves on the outside of the ingot show that a large number of the ingots have been poured in several times during the casting process (Stolz 2017a, 150).

The CT images already showed that within the eight groups, ten ingots were mainly arranged closely together and were tied into individual bundles when they were laid down (Fig. 6.1b). Occasionally, the individual bundles were even more clearly visible on the CT image than during the excavation process in the lab. To ensure that the position of the ingots is still traceable after uncovering and removal from their original position, all ingots were assigned numbers on the basis of the CT images. During uncovering, the ingots were always compared with the CT images, only then were the individual ingots finally numbered and finally removed. Groups 5, 6, 7, and 8 each consist of ten bundles of ten (one hundred ingots), group 2 consists of eight bundles of ten (eighty ingots). Only groups 3 (fifty-four pieces) and 4 (119 pieces), i.e. the two groups in the remaining block, and group 1 (141 pieces) differ in their irregular number of ingots. In group 1, one hundred ingots can be definitely assigned to ten bundles of ten, and forty-one ingots remain without definite assignment. Group 4 with 119 ingots consists of ten bundles of tens, a bundle of eight, and a bundle of eleven; group 3 consists of three bundles of ten and two bundles of twelve.

In addition to the location in the feature, organic remains (Fig. 6.8a) cover the bundles. These are unidentifiable untwisted bast, which has been preserved by the corrosion of the copper. The former lacing was partly traceable over several ingots, usually in the area of the ingot bends and ran straight or cross-shaped (Fig. 6.8b).

In addition to recording various characteristics, the individual weight of each uncovered ingot (621 pieces) was determined. On average, the ingots weigh 103 g. The lightest ingot weighs 76 g, the heaviest 161 g. The hoard of Oberding is thus characterized by a remarkably wide weight range, with the majority of the ingots

7 Fritz Moosleitner (1988, 43), for example, writes that 'the bar-shaped castings were probably still deformed in the hot state; by bending the ends open, one achieved the rib ingot shape'. Frauke Stein (1976, 40–41, 134 n. 134) also assumes that 'the ingots were deliberately bent before cooling to facilitate their transport in bundles'.

Table 6.1. Weight distribution of the weighed 621 ingots from groups 1, 2, 5–8. The ingots from group 3 and 4 were not exposed and remain in a rest block (LMU, Sabrina Kutscher 2017).

Weight Cluster (g)	**70–79**	**80–89**	**90–99**	**100–09**	**110–19**	**120–29**	**130–39**	**140–49**	**150–59**	**160–69**	
Weighted number of ingots: 621	1	62	219	176	96	40	15	8	3	1	66.37 kg
group											**Total weight of every group**
1 (141 items)	1	21	62	36	10	7	2	1		1	14.12 kg
2 (80 items)		2	27	22	20	3	2	3	1		8.59 kg
5 (100 items)		6	28	36	19	7	4				10.57 kg
6 (100 items)		15	39	25	10	9	1	1			10.15 kg
7 (100 items)		8	32	27	22	5	3	1	2		10.55 kg
8 (100 items		10	31	30	15	9	3	2			10.47 kg
3 (54 items)	No weight data (remaining block)										
4 (119 items)											
											Total weight hoard (projected): approx. 82 kg

weighing between 90 and 109 g (395 pieces), followed by ninety-six ingots weighing 110–19 g, and sixty-two pieces weighing 80–89 g (Table 6.1). That shows that the ingots were not precisely weighed. Nevertheless, they were aiming to produce ingots of similar weight (100 g ± 10–20 g), but also of similar length (30 cm ± 2–4 cm) — standard shapes. The combination of these characteristic types has shown that they are not interdependent, i.e. lighter ingots are not necessarily short and heavy ingots are not necessarily long.

Sixteen uncovered ingots as well as at least five ingots from groups 3 and 4 have sprues added subsequently. These can be divided into four variants.[8] Variant 1 (cuffs which completely enclose the body of the ingot) is represented nine times, variant 2 (elongated sprues in the middle part of the ingot) occurs twice, variant 3 ('cap-like' sprues on the tips of the ingot ends) is represented four times. An elongated sprue in the area of the bend on the outside of the ingot (variant 4) is represented only once (ingot 772). While variant 1 clearly fulfilled the purpose of a repair, the other variants may also have had the purpose of a 'beauty' correction (e.g. covering broken-off ingot tips or large, open blowholes) or residue recycling.

The X-rays show that in the case of seven ingots there was an antique fracture under the cuffs (variant 1) and that this fracture was obviously fixed by attaching a cuff (Fig. 6.9). In the case of another ingot with cuff, the fracture is externally visible due to the offset of the ingot ends themselves. In another ingot, the density of the cuff is too high, so that the ingot under the sprue is not visible on the X-ray image. The ingots with sprues weigh between 80 and 130 g. Due to the above-mentioned characteristics ('mending' of broken ingots), in my opinion the common interpretation of sprues as weight compensation does not always seem to be correct and certainly not the only 'plausible explanation' as Fritz Moosleitner (1988, 43 n. 16) once put it (Lenerz-de Wilde 1995, 229–327; 2002, 1–23; 2011, 177–98).[9]

On the long sides of four ingots between one and four notches (2–3 mm) were documented. In contrast to the sixteen ingots from Nonn (Berchtesgadener Land district), Alzgern (Altötting district), and Purkarec (Czech Republic), which are characterized by an extremely high lead content and differ from the other ingots, the ingots with notches from Oberding do not differ in terms of their metal composition from the other ingots of the Oberding group, which is rich in fahlore (Möslein 2015, 121, fig. 8).

Looking at the ingots of the individual bundles, it is clear that the ingot length was an important component in the composition of the individual bundles. In order to be able to stack the ingots compactly into each other,

8 The non-uncovered sprues were not counted at this point.

9 Václav Moucha (2005, 38) and Manfred Menke (1978–1979, 35) are also sceptical about the interpretation of the 'cuffs' as weight balancing.

Figure 6.9. So-called 'cuffs' — sprues enclosing the ingot body — served to fix fractures in the Oberding ingots, such as in ingot 588, and not as weight compensation (BLfD, Jörg Stolz 2016 and 2017).

it was necessary to use different lengths, whereby the differences in length are usually in the millimetre range.

Apparently, the total weight of the bundles of ten played a certain role. If one also considers the chronological predecessor type, which were ring ingots, it is noticeable that they were tied up in bundles of five and were also about twice as heavy as the Bermatingen type rib ingots, which were tied up in bundles of ten. This shows the targeted handling of the number ten and its multiples.

The comparison of the characteristics of the individual ingots shows that a unique position of the weight with regard to its standardization is not possible, but that the weight as well as the length, shape, completeness, and material were relevant (Dzbyński 2016, 71–92; Hensler 2014, 41–50).

The Copper Origin

First, the chemical composition of the copper ingots was determined by mobile X-ray fluorescence (XRF) analyses in the central laboratory of BLfD: in a first step, the ingots of group 1 were measured. Subsequently, the results were verified by twenty-three ED-XRF analyses and lead isotope analyses at the Curt-Engelhorn-Centre for Archaeometry in Mannheim and evaluated by Ernst Pernicka (Zentrallabor, BLfD, Björn Seewald; see Seewald 2017, 159–65). The results were largely consistent. Therefore, it was decided to apply mobile XRF analysis to the remaining uncovered ingots from the large block (Pernicka 2017, 169–79). Supplementary lead isotope and ED-XRF analyses were possible within the framework of the DFG project in order to clarify manufacturing issues on the one hand and to specify certain copper groups more precisely on the other.

Fahlore copper with nickel and a high content of arsenic, antimony, and silver, pure copper with low but approximately equal concentrations of arsenic and nickel, and a small number of ingots with a high nickel content could be determined (Krause and others 2018a, 169).

The pure copper probably comes from the eastern alpine deposits in the Mitterberg mining area. According to current research, large-scale mining there only began in the Middle Bronze Age. Fahlore copper, on the other hand, dominates mainly at the beginning of the Early Bronze Age and can be divided into two subspecies: with and without nickel. The so-called 'Singen copper' (with nickel) could originate from Slovakia, while the so-called 'Ösenringkupfer' (without nickel) probably originates from the Austrian Inn Valley (mining district between Schwaz and Brixlegg) (Bartelheim 2002; Junghans and others 1968; Junghans and Schröder 1974; Krause 2002, 44–59; Krause and others 2004, 4–21; Pernicka and Lutz 2015, 107–13; Schreiner 2007; Schubert and Pernicka 2013, 123–52). Both groups occur in all uncovered bundle groups and partly also mixed within bundles of ten. The joint depositing of different types of copper, especially ores from the Mitterberg region or the Slovakian Ore Mountains, which are important in the Middle Bronze Age, is an expression of new and important exchange relationships.

Works Cited

Ballmer, A. 2010. 'Zur Topologie des bronzezeitlichen Deponierens. Von der Handlungstheorie zur Raumanalyse', *Praehistorische Zeitschrift*, 85.1: 120–31.

Bankus, M. 2004. *Der Freisinger Domberg und sein Umland: Untersuchungen zur prähistorischen Besiedlung*, Freisinger Archäologische Forschungen, 1 (Rahden: Leidorf).

Bartelheim, M. 2002. 'Metallurgie und Gesellschaft in der Frühbronzezeit Mitteleuropas', in J. Müller (ed.), *Vom Endneolithikum zur Frühbronzezeit: Muster sozialen Wandels?*, Universitätsforschungen zur prähistorischen Archäologie, 90 (Bonn: Habelt), pp. 29–45.

Bátora, J. 2000. *Das Gräberfeld von Jelšovce/Slowakei: Ein Beitrag zur Frühbronzezeit im nordwestlichen Karpatenbecken*, Prähistorische Archäologie in Südosteuropa, 16.1 (Kiel: Oetker/Voges).

Blana, M., N. Reims, T. Fuchs, and M. Böhnel. 2016. '3D-Hochenergie-Computertomographie am Fraunhofer-Institut in Fürth', *Denkmalpflege Informationen*, 163: 48–50.

Brunn, W. A. von. 1995. *Bronzezeitliche Hortfunde*, I: *Die Hortfunde der frühen Bronzezeit aus Sachsen-Anhalt, Sachsen und Thüringen* (Berlin: Akademie-Verlag).

Chvojka, O., L. Jiráň, and M. Metlička (eds). 2017a. *Nové České Depoty Doby Bronzové: Hromadné nálezy kovových předmětů učiněné do roku 2013*, I (České Budějovice: Jihočeská univerzita v Českých Budějovicích).

———. (eds). 2017b. *Nové České Depoty Doby Bronzové. Hromadné nálezy kovových předmětů učiněné do roku 2013*, II (České Budějovice: Jihočeská univerzita v Českých Budějovicích).

David, W. 2015. 'Bronzedepotfunde der Früh- und Mittelbronzezeit in Ostbayern', in O. Chvojka, M. Chytráček, H. Gruber, L. Husty, J. Michálek, R. Sandner, K. Schmotz, and St. Traxler (eds), *Archäologische Arbeitsgemeinschaft Ostbayern/West- und Südböhmen/Oberösterreich*, Fines Transire, 24 (Rahden: Leidorf), pp. 87–112.

Dzbyński, A. 2016. 'Some Remarks on Number Concept Development in the Near East and Europe', *Archaeological Dialogues*, 23: 71–92.

Gasteiger, S. 2017. 'Der frühbronzezeitliche Spangenbarrenhort von Oberding – Abschluss eines Erfolgsprojektes', *Denkmalpflege Informationen*, 167: 60–61.

Hájek, L. 1954. 'Jižní Čechy ve starší době bronzové – Bohême méridionale à l'âge du bronze ancien', *Památky archeologické*, 45: 115–92.

Hansen, S., D. Neumann, and T. Vachta (eds). 2012. *Hort und Raum: Aktuelle Forschungen zu bronzezeitlichen Deponierungen in Mitteleuropa* (Berlin: De Gruyter).

Haßmann, H., A. Niemuth, M. Pahlow, B. Rasink, St Winghart, and F. W. Wulf. 2014. 'Der Goldhort von Gessel', in H. Meller, R. Risch, and E. Pernicka (eds), *Metalle der Macht – Frühes Gold und Silber: 6. Mitteldeutscher Archäologentag vom 17. bis 19. Oktober 2013 in Halle (Saale)*, II, Tagungen des Landesmuseums für Vorgeschichte Halle, 11.2 (Halle: LfA Sachsen-Anhalt), pp. 743–52.

Hensler, M. 2014. 'Überlegungen zur frühbronzezeitlichen Kupferdistribution am Beispiel der Ösenringe', in B. Nessel, I. Heske, and D. Brandherm (eds), *Ressourcen und Rohstoffe in der Bronzezeit: Nutzung – Distribution – Kontrolle*, Arbeitsberichte zur Bodendenkmalpflege in Brandenburg, 26 (Wünsdorf: Brandenburgisches Landesamt für Denkmalpflege und Archäologisches Landesmuseum), pp. 41–50.

Honig, P. 2008. *Studien zur bronzezeitlichen Keramikentwicklung am Beispiel der Siedlungskeramik der Windsheimer Bucht und des süddeutschen Donauraumes* (Büchenbach: Dr Faustus).

Innerhofer, F. 1997. 'Frühbronzezeitliche Barrenhorte – Die Schätze aus dem Boden kehren zurück', in A. Hänsel and B. Hänsel (eds), *Gaben an die Götter: Schätze der Bronzezeit Europas* (Berlin: Staatliche Museen Preußischer Kulturbesitz, Museum für Vor- u. Frühgeschichte), pp. 53–59.

Junghans, S., E. Sangmeister, and M. Schröder. 1968. *Kupfer und Bronze in frühen Metallzeiten Europas: Die Metallgruppen beim Stand von 12000 Analysen*, Studien zu den Anfängen der Metallurgie, 1–3 (Berlin: Mann).

Junghans, S., and M. Schröder. 1974. *Kupfer und Bronze in der frühen Metallzeit Europas*, Studien zu den Anfängen der Metallurgie, 2, 4 (Berlin: Mann).

Kopecky-Hermanns, B., and H. Krause. 2013. 'Ein Streifzug durch die Landschafts- und Erdgeschichte des Erdinger Landes', in J. Haberstroh, B. Päffgen, and W. Wagner (eds), *Stadt und Landkreis Erding: Ausflugsziele an Isar, Sempt, Isen und Vils* (Darmstadt: Theiss), pp. 14–28.

———. 2017. 'Geoarchäologische Spurensuche im Hortfund von Oberding – Bodenkunde auf kleinstem Raum', in Stadt Erding (ed.), *Spangenbarrenhort Oberding: Gebündelt und vergraben – ein rätselhaftes Kupferdepot der Frühbronzezeit*, Schriften, 2 (Erding: Museum Erding), pp. 64–77.

———. 2018. 'Geoarchäologische Spurensuche im frühbronzezeitlichen Spangenbarrenhort von Oberding – Bodenkundliche Untersuchungen zweier Blockbergungen', in St. Berg, E. Eckmeier, S. Linzen, L. Werther, and C. Zielhofer (eds), *Jahrestagung des Arbeitskreises Geoarchäologie: Anwendung und Weiterentwicklung geoarchäologischer Methoden und Konzepte in der archäologisch-bodendenkmalpflegerischen Praxis und Forschung; 04. bis 06. Mai 2018* <https://doi.org/10.22032/dbt.34521>.

Krause, R. 2002. 'Sozialstrukturen und Hierarchien – Überlegungen zur Frühbronzezeitlichen Metallurgiekette im süddeutschen Alpenvorland', in J. Müller (ed.), *Vom Endneolithikum zur Frühbronzezeit: Muster sozialen Wandels? Tagung Bamberg 14. bis 16. Juni 2001*, Universitätsforschungen zur prähistorischen Archäologie, 90 (Bonn: Habelt), pp. 44–59.

Krause, R., K. Oeggl, and E. Pernicka. 2004. 'Eine befestigte Burgsiedlung der Bronzezeit im Montafon, Vorarlberg. Interdisziplinäre Siedlungsforschung und Montanarchäologie in Bartholomäberg und im Silbertal', *Archäologie Österreichs*, 15.1: 4–21.

Krause, H., S. Kutscher, C. Metzner-Nebelsick, E. Pernicka, B. Seewald, and J. Stolz. 2018a. 'Europas größter Spangenbarrenhort: Der frühbronzezeitliche Kupferschatz von Oberding', in M. Wemhoff and M. M. Rind (eds), *Bewegte Zeiten: Archäologie in Deutschland* (Berlin: Imhof), pp. 167–69.

Krause, H., S. Kutscher, C. Metzner-Nebelsick, and E. Pernicka. 2018b. 'Kupferbarren – europäische Währung der Frühbronzezeit?', *Archäologie in Deutschland*, 3: 22–23.

Krenn-Leeb, A. 2008. 'Strategie und Strategem: Überlegungen zu Tradition, Innovation und Legitimation anhand der frühbronzezeitlichen Depotfunde in Österreich', in K. Schmotz (ed.), *Vorträge des 26. Niederbayerischen Archäologentages* (Rahden: Leidorf), pp. 163–96.

———. 2010. 'Ressource versus Ritual – Deponierungsstrategien der Frühbronzezeit in Österreich', in H. Meller and F. Bertemes (eds), *Der Griff nach den Sternen: Internationales Symposium in Halle (Saale) 16. bis 21. Februar 2005*, Tagungen des Landesmuseums für Vorgeschichte Halle, 5.1 (Halle: Landesamt für Denkmalpflege und Archäologie Sachsen-Anhalt), pp. 281–315.

Kutscher, S. 2017. 'Der frühbronzezeitliche Spangenbarrenhort von Oberding, Lkr. Erding. Erste Ergebnisse', *Archäologische Informationen*, 40: 423–34.

Later, C. 2017. 'Wissenschaftlicher Triathlon – Fachtagung, museale Präsentation und Begleitpublikation zum Hortfund von Oberding im Museum Erding', *Denkmalpflege Informationen*, 167: 83–85.

Leicht, J., and T. Stöckl. 2015. 'Ein Spangenbarrenhort der frühen Bronzezeit aus Oberding', *Das archäologische Jahr in Bayern*, 2014: 39–42.

Lenerz de Wilde, M. 1995. 'Prämonetäre Zahlungsmittel in der Kupfer- und Bronzezeit Mitteleuropas', *Fundberichte aus Baden-Württemberg*, 20: 229–327.

———. 2002. 'Bronzezeitliche Zahlungsmittel', *Mitteilungen der Anthropologischen Gesellschaft in Wien*, 132: 1–23.

———. 2011. 'Neue Ringbarrenhorte – Bronze als Wertträger (Prämonetäre Zahlungsmittel)', in U. L. Dietz and A. Jockenhövel (eds), *Bronzen im Spannungsfeld zwischen praktischer Nutzung und symbolischer Bedeutung: Beiträge zum internationalen Kolloquium am 9. und 10. Oktober 2008 in Münster*, Prähistorische Bronzefunde, 13 (Stuttgart: Steiner), pp. 177–98.

Massy, K. 2018. *Die Gräber der Frühbronzezeit im südlichen Bayern: Untersuchungen zu den Bestattungs- und Beigabensitten sowie Gräberfeldimanenten Strukturen*, Materialhefte zur bayerischen Archäologie, 107 (Kallmünz: Laßleben).

Menke, M. 1978–1979. *Studien zu den frühbronzezeitlichen Metalldepots Bayerns*, Jahresbericht der Bayerischen Bodendenkmalpflege, 19/20 (Munich: Habelt).

Moosleitner, F. 1988. 'Vier Spangenbarrendepots aus Obereching, Land Salzburg. Mit einem Beitrag von Hasso Moesta', *Germania*, 66: 29–67.

Moucha, V. 2005. *Hortfunde der frühen Bronzezeit in Böhmen* (Prague: Archeologický ústav AV CR).

Möslein, S. 1997. 'Frühbronzezeitliche Spangenbarrendepotfunde von Nonn, Stadt Bad Reichenhall, Landkreis Berchtesgadener Land, Oberbayern', *Das archäologische Jahr in Bayern*, 1996: 55–58.

———. 1998a. 'Überlegungen zur Bedeutung der Metallurgie in der Frühbronzezeit anlässlich des Ringbarrendepotfundes von Piding, Landkreis Berchtesgadener Land, Oberbayern', *Das archäologische Jahr in Bayern*, 1997: 73–74.

———. 1998b. 'Die Straubinger Gruppe der donauländischen Frühbronzezeit – Frühbronzezeitliche Keramik aus Südostbayern und ihre Bedeutung für die chronologische und regionale Gliederung der Frühen Bronzezeit in Südbayern', *Bericht der Bayerischen Bodendenkmalpflege*, 38: 37–106.

———. 1998–1999. 'Neue Depotfunde der älteren Bronzezeit aus dem oberbayerischen Alpenvorland. Ein Vorbericht', *Archäologie Österreichs Sonderausgabe*, 9–10: 69–77.

———. 2008. 'Frühbronzezeitliche Depotfunde im Alpenvorland – Neue Befunde', in K. Schmotz (ed.), *Vorträge des 26. Niederbayrischen Archäologentages* (Rahden: Leidorf), pp. 109–30.

———. 2015. 'Bronzezeitliche Hortfunde im ostbayerischen Alpenvorland', in O. Chvojka, M. Cgytráček, H. Gruber, L. Husty, J. Michálek, R. Sandner, K. Schmotz, and St. Traxler (eds), *Archäologische Arbeitsgemeinschaft Ostbayern / West- und Südböhmen / Oberösterreich*, Fines Transire, 24 (Rahden: Leidorf), pp. 113–22.

Nadler, M. 2003. 'Der Hortfund von Greding', *Das archäologische Jahr in Bayern*, 2002: 36–38.

Neugebauer, J.-W. 1977. 'Böheimkirchen. Monographie des namengebenden Fundortes der Böheimkirchnergruppe der Věteřovkultur', *Archaeologia Austriaca*, 61–62: 31–207.

Neumann, D. 2015. *Landschaften der Ritualisierung: Die Fundplätze kupfer- und bronzezeitlicher Metalldeponierungen zwischen Donau und Po* (Berlin: De Gruyter).

Pernicka, E. 2017. 'Untersuchungen zur Klassifikation und Herkunft der Spangenbarren von Oberding', in Stadt Erding (ed.), *Spangenbarrenhort Oberding. Gebündelt und vergraben – ein rätselhaftes Kupferdepot der Frühbronzezeit*, Schriften, 2 (Erding: Museum Erding), pp. 168–79.

Pernicka, E., and J. Lutz. 2015. 'Fahlerz- und Kupferkiesnutzung in der Bronze- und Eisenzeit', in T. Stöllner and K. Oeggl (eds), *Bergauf-Bergab: 10.000 Jahre Bergbau in den Ostalpen* (Bochum: Leidorf), pp. 107–13.

Press, E. 1953. *Vor- und Frühgeschichte des Landkreises Erding: Zwischen Sempt und Isen*, Heimatblätter des Erdinger Landkreises, 2 (Erding: Kreisverein für Heimatschutz und Denkmalpflege).

Ried, S. 1968. 'Oberding', *Bayerische Vorgeschichtsblätter*, 33: 178.

Rittershofer, K. F. 1983. *Der Hortfund von Bühl und seine Beziehungen* (Mainz: Von Zabern).

Ruckdeschel, W. 1978. *Die frühbronzezeitlichen Gräber Südbayerns: Ein Beitrag zur Kenntnis der Straubinger Kultur; Katalog – Tafeln – Karten* (Bonn: Habelt).

Schreiner, M. 2007. *Erzlagerstätten im Hrontal, Slowakei: Genese und prähistorische Nutzung*, Forschungen zur Archäometrie und Altertumswissenschaft, 3 (Rahden: Leidorf).

Schubert, M., and E. Pernicka. 2013. 'Die frühbronzezeitliche Kupferverarbeitung auf dem Buchberg im Unterinntal, Tirol', in K. Oeggl and V. Schaffer (eds), *Cuprum Tyrolense – 5550 Jahre Bergbau und Kupferverhüttung in Tirol* (Brixlegg: Edition Tirolense), pp. 123–52.

Seewald, B. 2017. 'Das Zentrallabor auf der Überholspur: 611 Spangenbarren in 8 Tagen analysiert! Metalluntersuchungen mittels mobiler Röntgenfluoreszensanalyse', in Stadt Erding (ed.), *Spangenbarrenhort Oberding: Gebündelt und vergraben – ein rätselhaftes Kupferdepot der Frühbronzezeit*, Schriften, 2 (Erding: Museum Erding), pp. 158–65.

Stadt Erding (ed.). 2017. *Spangenbarrenhort Oberding: Gebündelt und vergraben – ein rätselhaftes Kupferdepot der Frühbronzezeit*, Schriften, 2 (Erding: Museum Erding).

Stein, F. 1976. *Bronzezeitliche Hortfunde in Süddeutschland: Beiträge zur Interpretation einer Quellengattung*, Saarbrücker Beiträge zur Altertumskunde, 23 (Bonn: Habelt).

———. 1979. *Katalog der vorgeschichtlichen Hortfunde in Süddeutschland*, Saarbrücker Beiträge zur Altertumskunde, 24 (Bonn: Habelt).

Stolz, J. 2016. 'Computertomographie-Analysen an Blockbergungen aus Oberding – bisher größter bekannter Spangenbarrenhortfund', *Denkmalpflege Informationen*, 163: 51–53.

———. 2017a. 'Ungewöhnlich aber nicht unsinnig – die Herstellungstechnik von Spangenbarren', in Stadt Erding (ed.), *Spangenbarrenhort Oberding: Gebündelt und vergraben – ein rätselhaftes Kupferdepot der Frühbronzezeit*, Schriften, 2 (Erding: Museum Erding), pp. 149–57.

———. 2017b. 'Erste Nachweise des Dezimalsystems? Der frühbronzezeitliche Spangenbarrenhort von Oberding', *Restauro: Zeitschrift für Konservierung und Restaurierung*, 8: 14–18.

Wullinger, G. 2004. 'Eine Grube mit frühbronzezeitlicher Siedlungskeramik von Jellenkofen, Gde. Ergoldsbach, Lkr. Landshut', *Beiträge zur Archäologie in Niederbayern*, 2: 9–104.

THIBAUD POIGT AND COLINE RUIZ DARASSE

7. Numbers and Measures in the Iberian Peninsula during the Iron Age

Evidence from the Archaeological and Textual Records

ABSTRACT Studies on the use of numbers and measures during the European Iron Age are rare and poorly documented. The absence of written sources describing these practices partly explains this lack of interest. But it can also be credited to the dominant idea of a late importation of this knowledge into the Western world by the populations of the central and eastern Mediterranean.

This article aims to provide a brief and non-exhaustive overview of what we know about numbers and measures in one of the best examples of Late Prehistory in western Europe: the Iberian Peninsula. With the use of a homogeneous type of weights from the Late Bronze Age to the Roman period (a period of about one millennium) and the use of written numerical and metrological sequences, this region constitutes a very good indicator of the practices related to numbers and measures during a large period of time.

This study highlights the marginal role of numbers and measures in everyday activities and in written practices. Even though complex cognitive and practical counting and measuring instruments appeared very early in the Iberian Peninsula, they never seem to have been 'democratized' to the whole society.

KEYWORDS Late Prehistory, Palaeohispanic epigraphy, weights, metrology, counting

Introduction

Numbers and measures are not a well-known topic for Late Prehistoric Europe. Among the reasons for this silence, we can cite the lack of direct written sources or the difficulty of clearly identifying the measuring instruments. In this framework, the Iberian Iron Age (*c.* sixth–first centuries BC) is an exception with several studies focused on the metrological systems used and the existence of a local language known from hundreds of texts but still unintelligible.

In order to approach the uses of counting and measuring systems during the Iberian Iron Age, we can rely on different archaeological and epigraphical sources. The best evidence concerns the mass measurements which are relatively well studied based on the hundreds

Thibaud Poigt University of Bordeaux Montaigne. PhD, specialized in Bronze and Iron Ages weighing systems of western Europe; postdoctoral researcher UMR 5607 Ausonius and associate member UMR 5608 TRACES.

Coline Ruiz Darasse University of Bordeaux Montaigne. CNRS researcher, UMR 5607 Ausonius, specialist in linguistic and graphic contacts during the recent proto-history of the western Mediterranean, particularly in Palaeohispanic and Gaulish epigraphy.

Number and Measure in Prehistory: Gift, Exchange, and the Transfer of Goods, ed. by Philippe Della Casa and Aleksander Dzbyński, WSA, 6 (Turnhout, 2024), pp. 93–106.
BREPOLS PUBLISHERS DOI 10.1484/M.WSA-EB.5.137132

of scale weights identified in the Iberian Peninsula. Some information about this topic also comes from written sources and epigraphic marks. The broader issue of metrological systems is completed by one-off studies focused on modular architecture and vessels' volumes. The counting systems are more difficult to understand and are mainly deduced from the analysis of weights and from written evidence.

This article does not aim to resolve the complex issue of counting and measuring practices in the Iberian Peninsula during the Iron Age. Our objective is to offer a synthetic overview of the current knowledge on the subject. We will first deal with the archaeological and textual data concerning the measures and then we will deal with the question of numbers and counting in a second step.

To Measure and to Say the Measure

Measuring is an abstract process which is generally based on basic counting systems. Nonetheless, in the context discussed here and as for most of the non-literate ancient societies, we can rely on more data concerning the measures than counting. Most of the evidence comes from archaeological records and is the result of hypotheses and interpretations. The scope of metrology can be approached thanks to the measuring tools as well as the measured structures or artefacts. Metrology has a wide range of applications, from coins to huge architectural works. In the Iberian Peninsula the best evidence of metrological practices concerns the weights and scales, but some insights can be gathered about length and volume metrologies.

Length and Volume Metrology

We know truly little about the use of length and volume measurements in the daily life of the Iberians. Nevertheless, the results of some archaeological studies allow us to hypothesize that they were not totally unaware of the existence of such processes, at least during the end of the Iron Age (*c.* third–first centuries BC).

The works of Alain Badie and Pierre Moret (Badie and Moret 1997; 1998; Moret 2008) on several settlements of the eastern Iberian Peninsula such as the Puig de Sant Andreu (Ullastret, Gerona) or La Picola (Santa Pola, Alicante) proved the use of modular units in large architecture projects since at least the third century BC but maybe even before. The use of modules is a basic step towards the emergence of more complex metrological concepts (Dzbyński 2013, 27–34) and it does not specifically imply the existence of norms and length standards. However, such practices are good clues to conjecture that the Iberians, or at least the ones involved in the architectural works, have basic knowledge about numbers and measures. At the very least, they are used to duplicate basic length values through counting systems. The most common solution seems to be the combination of binary and ternary systems (Moret 2008, 93).

The topic of the volumes has been largely ignored for Iron Age Europe and the Iberian Peninsula is not an exception here. A recent study about the volumes of *c.* third–first-century BC amphorae shows that: unlike Italic or Punic ones, Iberian amphorae do not present a strong metrological structure. It seems that the potters' habits — the volume of clay he uses and the way he shapes it — are more important than seeking a given volumetric standard (Gorgues 2016; Gorgues and others in press).

So far, the data about length and volume measures in Iberian Peninsula is reduced and incomplete and more studies must be conducted to obtain a better understanding of this topic. The few clues we have suggest the Iberians could have basic knowledge about length measures and their use in architecture. Surprisingly, trading practices do not show any attempt at the normalization of the capacity of the amphorae in order to reduce the asymmetric information of the exchange (Gorgues 2016; Gorgues and others in press). The only way to understand better these concepts and their uses would be to identify the instruments which allow them: capacity table, measuring sticks, etc.

As we already said, the only well-known measurement instruments for Late Prehistoric Europe are the balance weights. In the Iberian Peninsula, we can distinguish two main discovery areas, presented separately here (Fig. 7.1).

The Eastern Coast Weighing System

The weighing systems of Iberian Peninsula are known by archaeological evidence (weights and scales) but also by several arithmetical sequences in Iberian texts. Even if we identify weights all across the Iberian Peninsula, most of the evidence for weighing practices comes from the eastern coast of the Iberian Peninsula, in a relatively small area encompassing the autonomous communities of Valencia and Murcia (Fig. 7.1). Most of the weights found in these areas have a relative homogeneous morphology and metrology, but they also do not bear numerical marks indicating their metrological value.

The Metrological Analysis of the Weights

The eastern Iberian weights have been identified since the 1930s (Ballester Tormo 1930) and have been regularly studied since (see Beltrán Villagrasa 1948; Cuadrado Díaz 1964; Fletcher Valls and Mata Parreño 1981; Fletcher

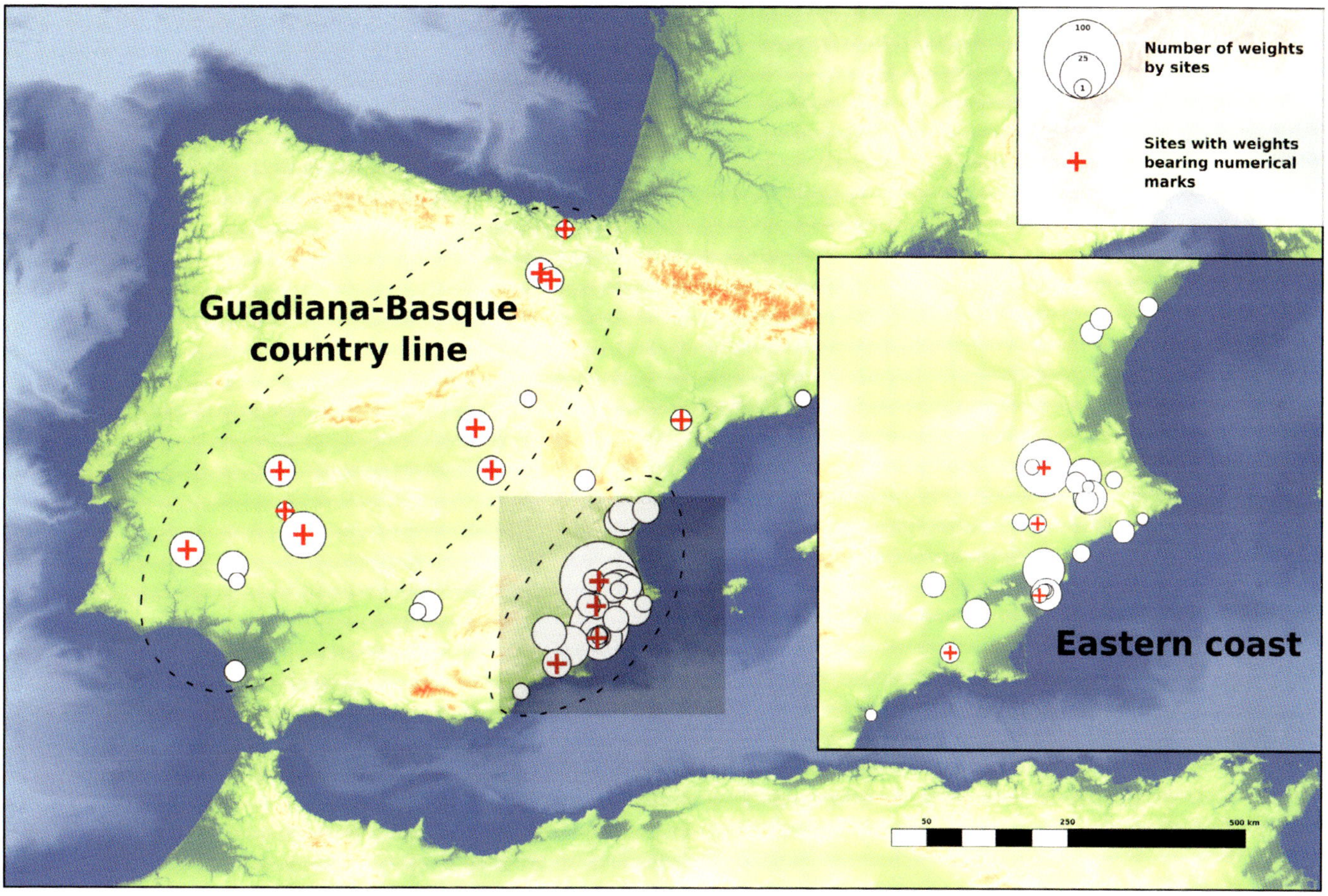

Figure 7.1. Distribution map of the Iron Age settlements with weights in the Iberian Peninsula depending on the number of artefacts and the presence of weights bearing numerical marks (CGIAR-CSI SRTM dataset; GIS: T. Poigt).

Valls and Silgo Gauche 1995; Grau Mira and Moratalla Jávega 2003–2004; Galilea Martínez 2004; Poigt 2015; Poigt and Delanaye 2020). They take the unusual form of perforated cylinders (proper cylinders, truncated cones, double cones, or discs) made of copper alloy or lead. The most ancient evidence of these shapes is dated to the Late Bronze Age in the south-west (Vilaça 2011) but they seem to be used in most parts of the Iberian Peninsula between the sixth and the first centuries BC, especially in the eastern part of it, where most of the weighing instruments come from (Fig. 7.1).

Several hypotheses have been put forward about the metrological system used by the Iberians. Most of them are based on a diffusionist model and the idea of a foreign origin of the weighing units (Beltrán Villagrasa 1948; Cuadrado Díaz 1964; Fletcher Valls and Mata Parreño 1981; Fletcher and Silgo Gauche 1995) which have been attributed mostly to the Phoenicians or the Greeks. Even if we know several complete sets of weights, the absence of any numerical mark on them — with few exceptions — left open the debate in the identification of their structuring unit. Depending on the error margin accepted, it becomes possible to make many assumptions on this topic.

However, the most recent works (Antunes 2017; Poigt 2022; Vilaça 2003; 2011) show that the weighing systems are known since the Bronze Age and that they are probably well embedded in the society — or at least some parts of it — since the fifth or fourth century BC. Consequently, the hypothesis of local metrological systems seems more adapted to the current state of the data than the diffusionist model. Even if some of these systems could originate from foreign units during the Late Bronze Age (see Pappa 2017; Poigt 2022, 253–59; Vilaça 2011, 155–64), it is clear that they should not be considered as anything other than local centuries later, during the Iron Age.

The current size of the corpus of Iberian weights (328 individuals) and the development of specific methods for metrological studies make it possible today to obtain a synthetic vision of Iberian weighing metrology. Furthermore, the Cosine Quantogram Analysis (CQA), combined with the Frequency Distribution Analysis (FDA), allows us to study large samples of metrolo-

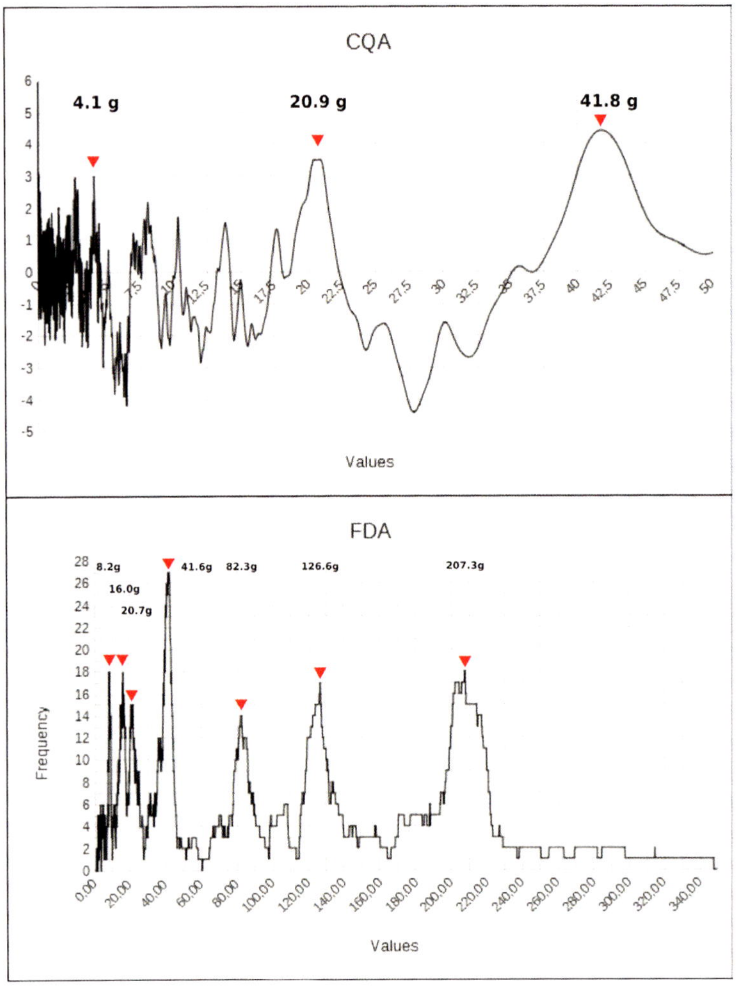

Figure 7.2. Cosine Quantogram (CQA) and Frequency Distribution analysis (FDA) of the eastern coast weights (311 ind. analysed) (processed with MetroTB v. 2.4; T. Poigt).

gical data in order to determine their most plausible structuring unit and the distribution of the different clusters of masses which are usually characteristic of a metrological construction (see Ialongo 2018; Ialongo and others 2018; Kendall 1974). For the eastern Iberian Peninsula, we can observe that the CQA highlights different *quanta* which could correspond to structuring units: 4.1 g, 20.9 g, and 41.8 g (Fig. 7.2). These units are actually very similar and correspond to different multiples of the same value (more or less 1 × 4.1 g, 5 × 4.1 g, and 10 × 4.1 g). The FDA clearly shows that the masses' clusters correspond to other multiples of the same values: around 8.2 g, 82.3 g, 126.6 g, and 207.3 g.

In a recent paper, Poigt (2015) suggested that the set of ten weights found in tomb 200 of the El Cigarralejo necropolis (Mula, Murcia), was structured around a unit of approximately 20.7 g. This unit has also been suggested by Galilea Martínez (2004) based on larger samples. The complete corpus seems to confirm this hypothesis with several sets built around the multiples 1: 2: 4: 5: 10 of this unit (Fig. 7.2). Nevertheless, other sets, composed of a reduced number of weights are probably built around the double of this 20.7 g unit, namely around 41–42 g. We can consider that this 'pair of units' (*c.* 21–42 g) represents the most common metrological structure for the Iberian area (see the complete demonstration in Poigt 2022, 360–70).

We can see through these results that the sets of weights are generally based on a quinary-decimal counting system even if it is generally possible to obtain higher values combining the weights. The most representative of this decimal system is probably the set from El Cigarralejo necropolis, already cited. It is composed of ten weights, the lightest one being the tenth of the unit and the heavier one being the tenfold of the unit (Poigt 2015).

The 'a-o-ki' System

In addition to the archaeological data, we can deduce some features of the weighing practices through some specific Iberian texts. Specifically, some written sequences contain what can be considered to be numerical information. In these sequences, one identifies vertical bars that precede or follow specific syllables of the Iberian script: **a**, **o**, and **ki** (Fig. 7.3). These sequences have been identified in at least nine Iberian texts (De Hoz 1981; 2011, 191–94; Oroz Arizcuren 1979a; 1979b) and the syllables **a**, **o**, **ki** always appear in the same order (Ferrer i Jané 2011, 101). From these texts, it is accepted that **a**, **o**, **ki** are metrological units and that the series of vertical bars are multipliers of each of them.

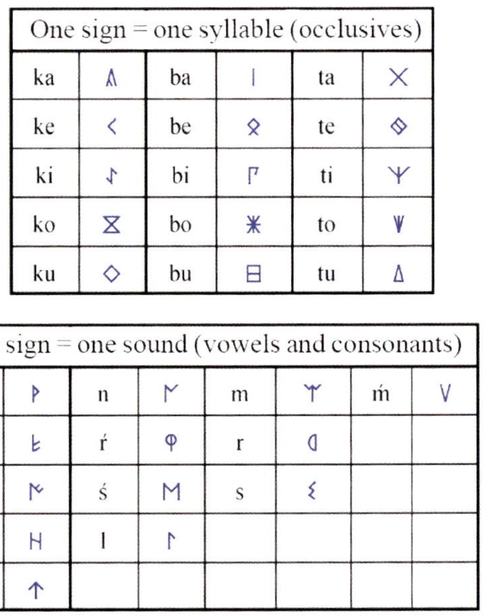

Figure 7.3. Eastern Iberian script (source: MLH-III, Hesperia; DAO: C. Ruiz Darasse).

Some scholars proposed that **a**, **o**, and **ki** are the abbreviations of the words **aba(ŕ)**, **ota(r)**, and **kita(r)** (Ferrer i Jané and Giral Royo 2007, 95; Fletcher Valls and Silgo Gauche 1995, 275; Orduña Aznar 2005, 496–99; Rodríguez Ramos 2005, 63; Vidal 2017). One of the main arguments for this hypothesis is that the word **kitar** is observed on Iberian coins, as well as the word **eta(r)**. It also strongly conditioned the idea that **e**, **a**, **o**, and **ki** correspond to value marks and more particularly that they indicate mass units (Ferrer i Jané 2011; Vidal 2017).

The written order between **a**, **o**, and **ki** and the use of the latter on coins, namely very light units, suggest that we can consider that **ki** is less than **o** which is less than **a**. Ferrer i Jané (2011) observed that the number of vertical bars attached to each sign never surpasses eight, which could indicate that the ratio between two units equals nine or more. If it was less than that, another sign could be written. In other words, if **a** equalled 8 × **o**, the sequences with eight vertical bars attached to **o** should be directly written **a**.

Vidal (2017) completed the data by using the archaeological evidence. As we said before, there are very few weights bearing numerical marks in the eastern part of the Iberian Peninsula and only two of them can be related to the Iberian syllabary. One of them, coming from Puig Castellar (Santa Coloma de Gramenet, Barcelona) weighs 423.78 g (Vidal 2017, 121) and bears the sequence **ustain** + **abaŕ** + **ar** + **ban** (BDH B.40.04, Fig. 7.4: a). It is heavier than most of the weights used in the Iberian Peninsula as we saw before and it bears what can be considered to be the larger metrological sign of the **a**-**o**-**ki** system. The second weight, coming from Puig de la Misericòrdia (Vinaròs, Castellón, BDH CS.02.01), weighs 41.29 g and bears the sign **o** (Fig. 7.4: b).

These two weights represent the only link between the written **a**-**o**-**ki** system and the metrological analysis inferred from the weights found in the archaeological record. But one can clearly see that the metrological connection between the two hypotheses can be easily highlighted. The 41.29 g weight from Puig de la Misericòrdia perfectly matches the unit of 41–42 g deduced from the analysis of the other weights one can call the unit **o**. The weight from Puig Castellar is approximately the tenfold of the one from Puig de la Misericòrdia and can be related to the unit **a**. This association implies that the ratio between **o** and **a** should be 1:10. This decimal relationship is perfectly corroborated by the FDA of the weights and can be considered as very plausible (Fig. 7.2).

We do not know any mention of the sign **ki** on weights and the coins bearing the word **kitar** do not match the decimal system, their average mass being 6.8 g (Vidal 2017, 124). Nevertheless, the metrological sequence of an unusual archaeological item found at La

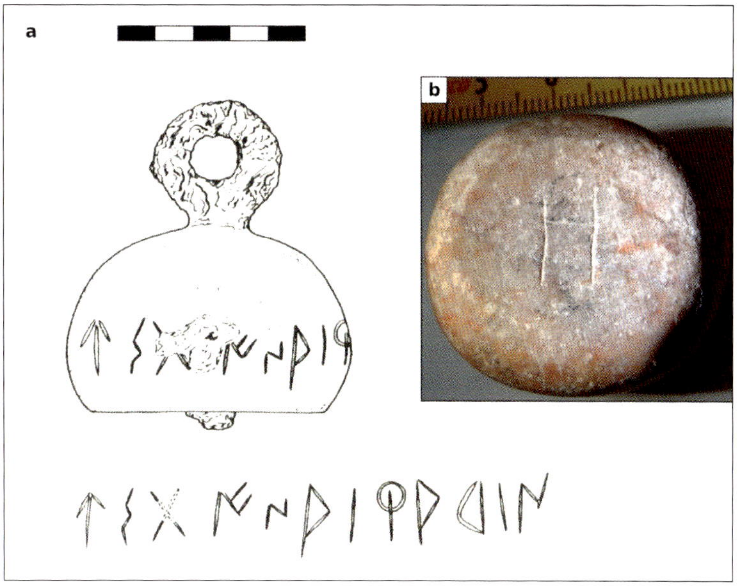

Figure 7.4. a) Weight from Santa Coloma de Gramanet (Barcelona) bearing the inscription '**ustain** + **abaŕ** + **ar** + **ban**' (after Sanmartí and others 1992, no. 153); b) weight from El Puig de la Misericòrdia (Vinaròs, Castellón) bearing the mark **o** (after Ferrer i Jané 2013, fig. 1).

Figure 7.5. Silver ware from La Granjuela and detail of the metrological sequence (after the notice produced by the Museo Arqueológico Nacional of Madrid; photographs: a) unknown; b) Ángel Martínez Levas).

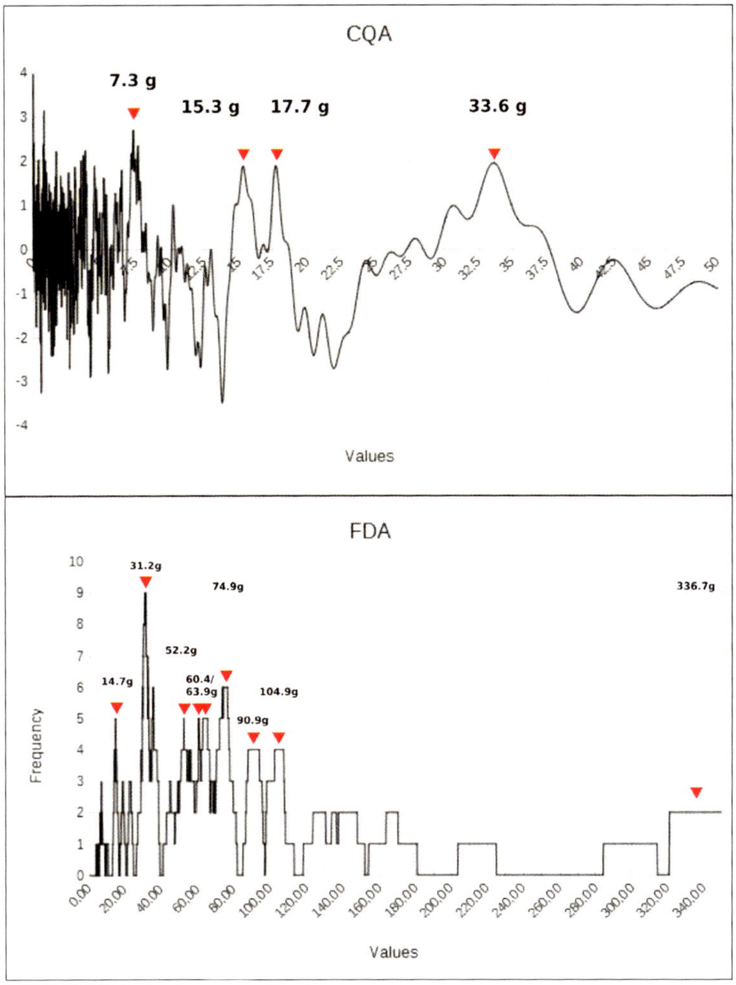

Figure 7.6. Cosine Quantogram (CQA) and Frequency Distribution analysis (FDA) of the Guadiana–Basque Country line weights (78 ind. analysed) (processed with MetroTB v. 2.4; T. Poigt).

Granjuela (El Alcornocal, Cordoba) has been largely used to conjecture about the **a-o-ki** system. This object artefact is a silver vase with no decorative element except for a short Iberian sentence and a metrological sequence from the **a-o-ki** system (BDH CO.01.01-MLH III, H.9.1) one can translate as **a** + 4 × **o** + 4 × **ki** (Fig. 7.5). The traditional hypothesis to explain this sequence is that it expresses the mass of the silver vase itself (Ferrer i Jané 2011, 108–21; Torija López 2003, 173; Tovar 1955). The reason for inscribing the metal mass of an object on itself during the Iron Age is extremely interesting and it has not been truly investigated so far, but from an arithmetical point of view, the hypothesis works perfectly. Indeed, if we consider an average value for **o** equalling 42 g and respecting a decimal ratio between the three units, then **a** equals 420 g and **ki** equals 4.2 g. The equation inscribed on the La Granjuela silver vase can be resolved as: 420 + 4 × 42 + 4 × 4.2 = 604.8 g. This result is remarkably close to the current mass of the silver vase: 606.1 g (Torija López 2003, 171).

The **a-o-ki** system presents several matches with the metrological systems deduced by the weights analysis. Nevertheless, the comparison is not completely convincing. Indeed, if the Iberian possessed a very theorized and normative weighing system with units having a written name, we can be surprised that only two weights out of several hundred actually bear these marks. Furthermore, as we said, the decimal relationship between **o(tar)** and **ki(tar)** is completely theoretical as the coins bearing the word **kitar** do not equal the tenth of a 42 g unit **o(tar)**.

The Guadiana–Basque Country Area Weighing System

The weights found outside the eastern coast of the Iberian Peninsula are mainly concentrated around a line joining the Guadiana Valley and the Basque Country with some discoveries as far as Portugal (Fig. 7.1). They are similar to the eastern ones in their general shape except that the double truncated cones are more frequent. The second distinguishing element is the presence of numerical marks on the weights. They are far more numerous than on the eastern coast and we can identify marks in every well-identified set of weights. These differences allow us to treat them separately as two distinct samples for the metrological analysis.

The Metrological Analysis of the Weights

If the studies concerning the eastern part of the Iberian Peninsula have been mostly focused on the research of a Greek unit in the Iberian sample, the ones dealing with the western and central parts were oriented towards the search for Phoenician relationships in the weighing systems (García-Bellido 2003; 2013; Pappa 2017; Vilaça 2003; 2011). This legacy is explained by the crucial role of the Phoenicians in the development of western Mediterranean exchanges and the evolution of the material cultures in the southern peninsula since the Late Bronze Age. Nevertheless, in the present state of knowledge, the most ancient weights found in the Iberian Peninsula appear to be locals and not Phoenicians (Poigt 2022, 247–49; Vilaça 2011, 155–58).

However, the CQA and FDA clearly show that the weights from the western areas are related to different metrological conceptions than the eastern ones (Figs 7.2 and 7.6). The CQA presents a high peak for a value of approximately 33 g, with very few alternative *quanta*. The FDA confirms the use of such a unit and some of its multiples, but it also appears that some clusters probably correspond to multiples of other units slightly lighter or heavier than 33 g. This dichotomy was observed very

Table 7.1. Synthetic table of the complete set of weights from the Guadiana–Basque Country line and their metrological characteristics; T. Poigt.

Site	Set	Chronology	No. of weights	Theoretical unit (g)	Biggest multiple	Lightest mass (g)	Theoretical of the biggest multiple (g)
Cancho Roano	Room E4 = H4	*c.* 5th c. BC	6	32.6	14	32.6	456.4
	Room N5	*c.* 450–400 BC	5	34.3–35.1	14	34.3–35.1	480.2–491.4
La Hoya	/	*c.* 400–350 BC	7	18.3	11	4.58	201.4
Haza de la Arca	/	*c.* 500–250 BC	6	34.7	20	23.1	694

Table 7.2. Synthetic table of the weights bearing numerical marks from Cancho Roano (Zalamea de la Serena, Badajoz) and the corresponding weighing units. Weight names and masses after Poigt 2022.

Weights from Cancho Roano	Current mass (g)	Numerical value marked	Corresponding unit (g)
CR-F	30.64	1	30.64
CR-K	61.95	2	30.98
CR-L	63.14	2	31.57
CR-Q	129.6	4	32.40
CR-S	161.79	5	32.36
CR-C	8.23	4 (divisor)	32.90
CR-M	65.77	2	32.89
CR-P	102.77	3	34.26
CR-N	70.24	2	35.12
CR-I	35.6	1	35.60
CR-X	72.64	2	36.32
CR-R	144.32	1	144.32
CR-U	298.31	2	149.16

early by García-Bellido (2003, 141–44) who conjectured that different units could be deduced from the Cancho Roano weights analysis. However, her final conclusions were that the weights corresponded to only one unit with a very wide deviation tolerance, up to 20 per cent (García-Bellido 2003, 141). This large deviation has been criticized by Grau Mira and Moratalla Jávega (2003–2004, 27) because it would annihilate the very basic function of a metrological system.

Analysing all the weights from the Guadiana–Basque Country line together, we can suppose that different systems have been used with probably no real attempt to match a specific metrological norm. For this reason, some sets display strong internal coherence and low deviation from the theoretical values, but they do not match each other (Table 7.1). We can probably consider that the whole area bases its weighing practices on a common metrological legacy — maybe inherited from the Late Bronze Age — but they use it with local goals and adaptations. Consequently, even if the weights used between the Guadiana Valley and the Basque Country during the Late Iron Age are roughly metrologically compatible, they were probably not dedicated to a common use at this interregional scale. This hypothesis is reinforced by the comparison between the metrological analysis results and the numerical marks on the weights.

The Numerical Marks

The numerical marks used in the Guadiana–Basque Country line are basic. They are mainly simple dots, lines, or semicircles repeated on the weights depending on the multiple represented. In some cases, one can also observe that the number of marks may express a fraction. For example, at Cancho Roano (Zalamea de la Serena, Badajoz), the marks are circles containing the value of the multiple. But in at least one case, a weight bears four circles organized as a square which express a fraction: the fourth of the unit.

In the whole peninsula, these marks never overpass the number five which could be understood as the use of a quinary counting system. Among the twenty-three weights found at Cancho Roano, at least thirteen of them bear a mark (García-Bellido 2003). For eleven of them, it allows us to determine a unit between 30.6 g and 36.3 g (Table 7.2). Two of them correspond to a heavier unit of approximately 144.3–149.2 g — even if we cannot see any difference in the shape of the marks. This unit can probably be understood as the fivefold of the lighter one (*c.* 31 g) which corroborates the idea of a quinary counting system. The same can be said with other complete sets of weights ended by a fivefold multiple such as the ones from Llano de la Horca (Santorcaz, Madrid; Ruiz Zapatero 2012, 358) and La Custodia (Viana, Navarra; Labeaga Mendiola 1999, 132). The set from La Haza del Arca (Uclés, Cuenca), by contrast, is comprised of six weights with the heavier corresponding to the tenfold unit (Lorrio 2007, 255–56).

Besides allowing considerations about the counting system used, the numerical marks offer a good base to highlight the diversity of metrological units structuring the weighing practices in the Guadiana–Basque Country line. If we consider each weight bearing the mark 'one', we find in this area several weighing units. We can firstly observe 'light' units: a first one equalling approximately 15.3–15.4 g identified at Munoaundi (Azkoitia-Azpeitia, Basque Country; Peñalver and San Jose 2011, 129) and La Custodia (Viana, Navarra) and another of 17–18.3 g at Llano de la Horca (Santorcaz, Madrid), La Hoya (Guardia, Basque Country), and La Custodia. As for the pair of units 21–42 g deduced from the analysis of the eastern weights, these light units seem to have been doubled in the archaeological record. At Azougada (Moura, Alentejo, Portugal) a unit of around 31.3 g is used, approximately the twofold of the 15.3–4 g unit (Antunes 2017, 911–17), and a 34.7 g unit is used at Haza de la Arca (Uclés, Cuenca) which could be the double of the 17–18.3 g unit. As we said before, a wide range of units between 30.6 and 36.3 g are induced by the observation of the Cancho Roano weights. The 'heavy' unit of 144.3–149.2 g is only identified as such at Cancho Roano (García-Bellido 2003, 141–44).

These results let us think that we can find the same kind of pair-units structure in the whole peninsula during the Late Iron Age but depending on several main weighing units (*c.* 15 g, *c.* 17–18 g, *c.* 21 g) and based on different visual proxies indicating the metrological value of the weights (several numerical marks, Iberian written symbols, location in the set of weights, etc.). The **a-o-ki** system seems to have no existence in the Guadiana–Basque Country line, even if the discovery of the silver vase from La Granjuela is geographically closer to this area than the eastern coast and the script used is that of the north-east. On the contrary, the use of visual numerical marks is mostly restricted to the Guadiana–Basque Country line except for a few specific weights.

To Count, to Say, and to Write Numbers

The Decimal Counting System: The Weights Evidence

The study of Iberian Peninsula weights show that the whole area shares common characteristics in their metrological practices. First, we can see that the structuring units are relatively heavy — between 15 and 36 g for most of them — and they are mostly used with multiples rather than fractions. Furthermore, even if we cannot talk about metrological norms, we see that there exist standard units which often appear as pairs in the archaeological record sharing a ratio 1:2.

Another shared feature seems to be the use of a quinary-decimal counting system. This practice appears clearly in the building of weighing unit multiples across the Iberian Peninsula. The fractions are generally simpler and are based on basic subdivisions: halves, thirds, and quarters; even if we can occasionally observe a more sophisticated system using fractions based on five or ten (Poigt 2015). The creation of submultiple weights is probably more a matter of feasibility than the use of a specific weighing system. Even today, we barely use the tenth fractions even if our whole counting system is decimal.

It is also interesting to see that there is probably a correlation between the choice of the counting system — quinary or decimal — and the unit used among the pair. The light units (*c.* 15 g, 18 g, or 21 g) are generally used with a decimal construction of the weights set while the quinary system is more frequent with the heavier units (*c.* 30 g, 36 g, or 42 g). As we saw before, this binary approach of measures can also be suggested by the study of the modular architecture (Badie and Moret 1998; Moret 2008). Even if it is less clear, we can observe that few Iberian amphorae have a capacity equalling the half of the main capacities cluster (Gorgues and others in press).

Iberian Numbers and Basque Numbers

In addition to weighing systems and metrological marks, there are also artefacts in the Iberian Peninsula that refer more directly to the practice of counting, in particular some Palaeohispanic texts. Traditionally the Iberian language is interpreted as a language with an economic vocation. For this reason, most long texts are understood as documents with a commercial purpose.

Certain types of texts seem to carry numerical indications or even sometimes deliver numerical sequences written in full letters. Linguistic and lexical hypotheses have been proposed for these sequences.

The starting hypothesis is an equivalence between the lexical form of numbers in the Basque language and certain sequences read in Iberian-language texts. In other words, the names of Basque numbers would be the same as those of Iberian numbers, written in plain text. This equivalence is therefore in accord with the so-called Vasco-Iberist theory (for a recent state of the art on this theory, see Orduña Aznar 2019, 219–40. For a critical review about numbers see Lakarra 2010 and the response in Orduña Aznar 2011). The hypothesis was already proposed in 1985 by one of the leading figures in Vasco-Iberism, Luis Michelena, and was taken up and developed by Eduardo Orduña Aznar in 2005, followed by the work of Joan Ferrer i Jané in the next decade.

The hypothesis starts from a sequence of Iberian language read on a lead found at Ullastret (BDH GI.15.04, face A l. 2) **borste.abaŕkeborste** (Orduña Aznar 2005). Its structure is complex. It has been analysed based both on the Basque system and on the way of composing numbers for example in German, quite a different language from Iberian (for an overview on how to say numbers in the languages of the Mediterranean area, see Valeri 1999). Based on the equivalence between Iberian **bors(te)** and Basque *borstz/bost* for the number five, Orduña Aznar (2005, 496) proposes to see in Iberian **abaŕ** (Basque *[h]amar*) the name of the ten and advances 'fifty-five' as a translation for the Ullastret sequence.

On this very point, we can remark that, considering the **a-o-ki** system, **a** could correspond to **abaŕ** which seems equivalent to the tenfold of **o**.

Subsequently, other sequences could be identified (see Table 7.3, according to Ferrer i Jané 2009, 472, table 10). These propositions are then based on lexical equivalences but also on combinations of these elements with each other, for example through the 'conjunction' **ke**. On this basis, it has thus been possible to propose names for numbers beyond twenty, for example thirty, as on the ostracon of Can Vedell (BDH B.22.02) where **oŕkeiŕur**[/ **oŕkeibaŕban**[equivalent according to the reading of Ferrer i Jané (2011, 126) at 23 and 31, or eighty, by contracting **la(ur)** with **(oŕ)kei** in order to compose a **lakei**, which has been spotted on several occasions (including perhaps once at the Caserne Niel in Toulouse? cf. Moret and others 2015).

Thus, the comparationist starting point (with Basque) is completed and seems corroborated by an internal analysis of the Iberian data.

In long texts whose meaning remains completely impenetrable for the moment, such as the lead found at Ensérune (BDH HER.02.373), a welcome light is shed when a possible numeral element is identified there, for example **śisbi** on line 1 of this lead. Moncunill Martí (2007, 285) and Orduña Aznar (2005) consider it to be a numeral in non-digital script (a number written in letters). But this identification influences the interpretation of the nature of the document produced: by analogy with the only other lead related to the Iberian world we know of and that we can understand, namely the Greek lead of Pech Maho, it is considered to be a commercial lead. However, there is no other sequence that could come close to figures elsewhere on this same lead, either in a graphic or lexical form.

Table 7.3. Summary table of possible notations of the numbers in letters after Ferrer i Jané 2009, table 10; DAO: C. Ruiz Darasse.

Value	Iberian	Mark of value	Basque	Possible abreviation
1/2	**erdi**	**erder**	*erdi*	**e**
1	**ban**	**ban**	*bat*	**ban/ba**
2	**bi(n)**		*bi*	**bi**
3	**irur**		*(h)iru(r)*	**i**
4	**lau(r)**	**l?**	*lau(r)*	**l**
5	**bors(te)**		*bortz/bost*	**bo**
6	**śei**	**śeŕki**	*sei*	**ś**
7	**śisbi**		*zazpi*	**ś**
8	**sorse**		*zortzi*	**s**
10	**(a)baŕ**		*(h)amar*	**a**
20	**oŕkei**		*(h)ogei*	**o**

Finally, the system established by Orduña Aznar and Ferrer i Jané also seems to function in abbreviated form, if we refer to the marks on the coins (Ferrer i Jané 2009, 451–79, which takes up some aspects already suggested by Villaronga 1973). Indeed, several signs appear in addition to the monetary legends on the Iberian coins to which the authors have assigned different values (**e** = **eta** = (bronze) unit; **ba** = **ban** = 1; two dashes = **erder** = ½; one dash = ? = ¼). In the table (Table 7.3), the last column is our responsibility and is not proposed either by Ferrer i Jané or Orduña Aznar: it is intended to see to what extent certain syllabograms recurrently present on archaeological objects (mostly ceramics) could also have a numerical value, considering them as the abbreviations or acronyms of the names of the numbers. But the marks that are found in a very repetitive manner on ceramic objects are too varied to be reduced to these initials. It is therefore difficult to think that certain syllabograms were used as numbers, according to an acronymic logic.

Numancia BDH SO.01.13 ©ISC, Hesperia, CC BY-NC 3.0						
Possible values in Western Celtiberian script (cf. Rodríguez Ramos 2004)	?	**ke** 3 or **to** 3	**ta** 1	**r** 6	**ki** 1	**n** 1
Mas de Moreno (Foz) ©C. Sacilotto						
Possible values in Northeastern Iberian script (cf. Rodríguez Ramos 2004)	**i** 1	**be** 1	**ta** 1	**ti** 5	**bi** 1	**ḿ** 1

Figure 7.7. Comparison of the dice of Numancia and Foz (DAO: C. Ruiz Darasse).

Nevertheless, the whole system seems to be consistent except for a few remarks (see below in the Discussion).

Numbers on Dices

To end with the idea of an acronymic logic, we have to complete this series of suggestions concerning figures in the Iberian world with two documents that do not immediately belong to our corpus of study: two dice.

Several dice are known in the Iberian Peninsula, probably with a playful or mantic dimension, which only bear marks similar to those known elsewhere in the Etruscan or Roman world. Others, either in ceramic in Sepúlveda (Blanco García 2004) or in stone in Calagurris (Cinca Martínez and Ballester 1998), do not have any signs on their faces but floral, geometric, or even anthropomorphic forms. For an overview of all the dice found in the Iberian Peninsula, their distribution, and study, see Blasco Martín 2016.

But two dice from the Iberian Peninsula are distinguished by the use of signs that can be interpreted as syllabograms on each side: a dice found in Numance during the great excavations of Adolph Schulten (Arlegui and Ballester 1997; Schulten 1931, pl. 37 B3) and another from the Mas de Moreno (Foz Calanda, Teruel) found during the last excavation campaign in 2015 directed by Alexis Gorgues and José Antonio Benavente Serrano (Gorgues and others 2021). The one from Numance, kept at the Museo Numantino inv. 12.795, is made of stone and the second, kept at the CIBA in Alcañiz, is made of ceramic; both can be dated to the first century BC, even though the Numantine die has no precise archaeological context. Each of the faces of these dice are incised with what appears to be a syllabogram belonging to a Palaeohispanic script (Fig. 7.7: for palaeographic references see Rodríguez Ramos 2004). Only two signs are exactly the same from one die to the other: <**ta**> (cross-shaped) and <**n**> or <**l**> (chevron-shaped).

It is possible that the syllabograms present in Numance and Foz correspond to a numerical system, presented in the form of an acronym, but these 'initials' do not correspond to the lexical form of those we have seen for Iberian numbers (see Table 7.3), nor to their possible abbreviations, whatever order is chosen.

What can we deduce from this?

There is no evidence that these dice had a numerical function. However, there are many examples of dice being used in playful settings, even in antiquity, without the elements represented on each side being numbers.

It is also possible that the names of the numbers are not identical throughout the Iberian world or that the Iberian world, in its multiplicity, did not have a uniform numerical system. The numbers and their names or even the symbol to represent them (but not their values) may have varied from one region to another. Unfortunately, we cannot compare with the Celtiberian domain, because no texts with number names have come up to date. The only indications referring to numbers are to be deduced from lexical or anthroponymic structures (cf. **tirikantam** or **kuintiaku**, Colera 2019, 235–36).

Finally, as Blasco Martín (2016, 250) suggests for a dice found in Oliete dated from the first third of the first century BC (quoting Manniez 2010, 20), we cannot rule out the possibility of a search for originality on the part of the craftsman who shaped the dice.

Discussion

Through our analysis, we highlighted that there is not one but several metrological systems in the Iberian Peninsula during the Iron Age, specifically applied to the weighing measures. The most representative is the one used on the eastern coast but at least one other main system (probably based on several different metrological practices) is used in the hinterland and the Meseta.

Several evidences suggest that the graphic **a-o-ki** system corresponds to the written equivalent of the metrological weighing system used on the eastern coast. Nevertheless, the metrological sentences using it mostly appear inside larger texts and only on two late weights out of more than three hundred identified.

It is also important to notice that the metrology does not seem embedded in all spheres of the society. The Iberian people do not involve the metrological tools in all of their daily practices. When the study of the discovery contexts of the scale weights allows it, one can even suggest that the weighing corresponds to an elitarian practice (see Poigt 2022, 316–34, 390–92).

Iberians almost certainly use only one counting system, mostly decimal, with different practical uses for weighing. Our understanding may also be blurred by a binary conception of measure practices which leads to the observation of 'pair-units'.

As for the written notation of this counting system, the current state of research does not allow any definitive conclusion to be drawn despite the proposition of several lexical hypotheses, based on linguistic comparisons. Furthermore, so far, neither the texts nor the discovery of the dice have assured the use of written numbers by the Iberian people.

Indeed, no written mark is identified that could correspond to what is usually known for the notation of numbers (vertical bars for a unit, acronym principle like the pi/penta in Greek for five, or recurrences of simple characters like the Latin V and X). It is an unusual case study that partly explains why the numbers could be written in words in some Iberian texts in the absence of specific symbols.

To give but one example, we can stress the **ban** sequence, which is attested in many texts but without any explicit numerical meaning. It is possible that this element corresponds to an indeterminate article, just as in French *un* (Orduña Aznar 2011, 136), and always written in full. But in Iberian **ban** begins precisely with a simple vertical line, and this very vertical bar (equivalent to the syllabogram <**ba**>) can also be found and used as a counting mark, without the use of full written letters. In some lead plaques with 'numerical' sequences like the one found in Los Villares (BDH V.07.01), we can observe serials of vertical lines (= the syllabogram **ba**), up to twenty of them. It would then be good to understand why some complete sequences are in letters (such as Ullastret's **borste.abaŕkeborste**) and others are not simplified according to a principle system when counting. Is it the nature of the elements involved?

This begs the question, in what context do we prefer to write numbers in letters rather than create numerical symbols? This point is purely conventional, and therefore cultural. The remarks we will be able to make here are therefore exclusive to the European, French, and modern field where it is possible to write in both letters and numbers. Numbers written in letters appear in legal texts such as contracts or notarial documents, where the numbers must be written in letters in order to avoid subsequent falsification. However, there are other cases outside the commercial or legislative context. There are other conventions (beginning of sentence, numbers used as nouns, historical periods, duration, distance, or unit of measurement not followed by a symbol) in which numbers are conventionally written in letters. Sequences from the Iberian texts could be considered in the latter case.

Traditionally, the Iberian language is considered a language with a commercial and economic purpose (De Hoz 2011, 445). One would expect therefore graphic practices to integrate more deeply the numerical data and on a variety of media and materials. Nevertheless, things are not uniform: counts (with sequences of vertical bars) are much more frequent on metal elements (lead plaques, silver ware, coins) than on ceramics.

One can ask if such a situation is not the result of a late adoption of writing coming after a long oral practice of counting. The graphic system not having the appropriate symbols for representing numbers, it only transcripts 'in extenso' the name of the numbers traditionally used by the Iberian people, probably for centuries. Adding to that, one can remember that Iberian is not an Indo-European language and, consequently, the usual references for conceptualizing numbers are useless here.

If we accept the hypotheses of an early conceptualization of numbers based on oral practices only, it seems possible that, once some Iberians begin to use writing systems, borrowed from foreign peoples, numbers are just transcribed as any other word. It is interesting to note that such a situation would be the opposite of what occurred in the Near East where the writing systems seem to be developed firstly with a strong numerical basis.

Conclusion

Numbers and measures are a large topic known only by a heterogeneous corpus for the Iberian Peninsula during the Iron Age. Regardless of the strong convergences in the use of metrological system (types of balance weights, decimal counting system, use of 'pair-units', regional embedding of the units) and a large use of the Iberian writing system, particularly in settlements where weights have been discovered, the correlation between the two practices is never systematic.

The use of a weighing metrological system does not imply the writing of this system and the writing is only used marginally for metrological purposes.

The existence of metrological sequences, composed of multipliers and units, shows that there is a real practice of arithmetic in the Iberian Peninsula during the Late Iron Age. However, even if we still cannot understand the Iberian language, it seems impossible to properly identify numbers as graphic elements either in texts or on dice. This set of observations suggests that the written medium is not of primary importance in the implementation of arithmetic. This does not exclude the presence of counts on certain supports in the form of sequences of vertical bars including metrological sequences. Writing could have been used to transcribe the name of numbers more than to develop a specific notation of the value of numbers. As we proposed, this could be the consequence of a late adoption of the written scripts in comparison to the ancient practice of counting and weighing.

This brief overview of numbers and measurements in the Iberian Peninsula during the Iron Age is not intended to be considered as exhaustive. It must be completed by several studies such as, for example, the metrological marks of amphorae discovered in the settlement of Vieille-Toulouse (Haute-Garonne, France), which correspond to an Iberian script used in a multicultural context even if they are more recent than most of the artefacts studied here. Nevertheless, this general synthesis has highlighted the complexity of the relationships between counting systems, weighing systems, and graphic practices.

Works Cited

Antunes, A. S. 2017. 'A Azougada (Moura) e o sistema metrológico da Idade do Ferro Pós-Orientalizante do Baixo e Médio Guadiana', in J. M. Arnaud and A. Martins (eds), *Arqueologia em Portugal 2017: estado da questão* (Lisbon: Associação dos arqueólogos portugueses), pp. 907–28.

Arlegui, M., and X. Ballester. 1997. 'El dado numantino', *Kalathos*, 16: 213–22.

Badie, A., and P. Moret. 1997. 'Métrologie et organisation modulaire de l'espace au vème siècle av. J.-C. sur le site ibérique de la Picola (Santa Pola, Alicante)', *Pallas*, 46: 31–49.

———. 1998. 'Metrologia y arquitectura modular en el puerto de la Picola (Santa Pola Alicante) al final del siglo V a. C.', *Archivo español de arqueología*, 71.177–78: 53–61.

Ballester Tormo, I. 1930. 'Los ponderales Ibéricos de tipo covaltino', in *Comunicaciones al IV congreso internacional de arqueología* (Valencia: Tipografía Moderna), pp. 1–25.

Beltrán Villagrasa, P. 1948. 'El "ponderarium" de Covalta y la "Mina cobaltina"', *Saitabi*, 6.28: 131–37.

Blanco García, J. F. 2004. 'Pieza cúbica celtibérica de arcilla hallada en Sepúlveda', *Cuadernos de prehistoria y arqueología*, 30: 131–40.

Blasco Martín, M. 2016. 'Dados y fichas de la Edad del Hierro en la Península Ibérica', *Archivo de prehistoria levantina*, 31: 241–60.

Cinca Martínez, J. L., and X. Ballester. 1998. 'El dado calagurritano', *Kalakorikos*, 3: 233–38.

Colera, C. 2019. *Lengua y epigrafía celtibéricas* (Zaragoza: Monografías de filología griega).

Cuadrado Díaz, E. 1964. 'Sobre ponderales ibéricos', in *VIII Congreso nacional de arqueología, Sevilla-Malaga, 1963* (Zaragoza: Universidad de Zaragoza, Seminario de arqueología), pp. 339–52.

De Hoz, J. 1981. 'Algunas precisiones sobre textos metrológicos ibéricos', *Archivo de prehistoria levantina*, 26: 475–86.

———. 2011. *Historia lingüística de la Península Ibérica en la Antigüedad*, II: *El mundo ibérico prerromano y la indoeuropeización* (Madrid: Manuales y anejos de Emérita).

Dzbyński, A. 2013. *The Power of the Line: Metaphor, Number and Material Culture in European Prehistory* (Newcastle upon Tyne: Cambridge Scholars).
Ferrer i Jané, J. 2009. 'El sistema de numerales ibérico: avances de su conocimiento', *Palaeohispánica*, 9: 451–79.
———. 2011. 'Sistemas metrológicos en textos ibéricos (1): del cuenco de La Granjuela al plomo de La Bastida', *Estudios de lenguas y epigrafía antiguas*, 11: 99–130.
———. 2013. 'A propòsit d'un pes de pedra ibèric del Puig de la Misericòrdia (Vinaròs) de 41 gr amb la marca metrològica "o"', *Cuadernos de prehistoria y arqueología castellonenses*, 31: 137–47.
Ferrer i Jané, J., and F. Giral Royo. 2007. 'A propósito de un semis de "ildirda" con leyenda "erder": marcas de valor léxicas sobre monedas ibéricas', *Palaeohispánica*, 7: 83–99.
Fletcher Valls, D., and C. Mata Parreño. 1981. 'Aportación al conocimiento de los ponderales ibéricos', *Saguntum*, 16: 165–76.
Fletcher Valls, D., and L. Silgo Gauche. 1995. 'De nuevo sobre ponderales ibéricos', *Verdolay*, 7: 271–75.
Galilea Martínez, F. 2004. 'Los sistemas de pesos en la Península Ibérica durante la II Edad del Hierro', *Estudios de arqueología alavesa*, 21: 209–52.
García-Bellido, M. P. 2003. 'Los ponderales y sus funciones económica y religiosa', in S. Celestino Pérez (ed.), *Cancho Roano IX: Los materiales arqueológicos*, II (Merida: Junta de Extremadura, Consejería de Cultura), pp. 125–56.
———. 2013. 'Los sistemas ponderales en el mundo púnico de Iberia e Ibiza', in B. Costa and J. H. Fernández (eds), *La moneda y su papel en las sociedades fenicio-púnicas (Eivissa, 2012): XXVII jornadas de arqueología fenicio-púnica* (Eivissa: Museu Arqueologic d'Eivissa i Formentera), pp. 35–60.
Gorgues, A. 2016. 'Trade in a Liminal Zone: Commercial Encounter and Transformation in the Iron Age North-West Mediterranean', in I. Armit, H. Potrebica, M. Črešnar, P. Mason, and L. Büster (eds), *Cultural Encounters in Iron Age* (Budapest: Archaeolingua), pp. 167–210.
Gorgues, A., F. Comte, and T. Poigt. in press. 'True Lies: The Assessment of Volume and the Reliability of Information in the Ancient Western Mediterranean Seaborne Trade (c. 250–50 BCE)', in N. Badoud and C. Doyen (eds), *Un marché commun dans l'Antiquité?* (Bâle: Schwabe).
Gorgues, A., C. Ruiz Darasse, and C. Sacilotto. 2021. 'Un dé ibérique et son contexte: l'objet de l'atelier de potiers du Mas de Moreno (Foz-Calanda, Espagne)', *Palaeohispanica*, 21: 495–511.
Grau Mira, I., and J. Moratalla Jávega. 2003–2004. 'La regulación del peso en la contestania ibérica. Contribución al estudio formal y metrológico de las pesas de balanza', *Anales de prehistoria y arqueología*, 19–20: 25–54.
Ialongo, N. 2018. 'The Earliest Balance Weights in the West: Towards an Independent Metrology for Bronze Age Europe', *Cambridge Archaeological Journal*, 29.1: 103–24.
Ialongo, N., A. Vacca, and L. Peyronel. 2018. 'Breaking down the Bullion. The Compliance of Bullion-Currencies with Official Weight-Systems in a Case-Study from the Ancient Near East', *Journal of Archaeological Science*, 91: 20–32.
Kendall, D. G. 1974. 'Hunting Quanta', *Philosophical Transactions of the Royal Society of London: Mathematical and Physical Sciences*, 276.1257: 231–66.
Labeaga Mendiola, J. C. 1999. *La Custodia, Viana: Vareia de los Berones* (Pamplona: Trabajos de Arqueología Navarra).
Lorrio, A. 2007. 'Historiografía y nuevas interpretaciones: la necrópolis de la Edad del Hierro de Haza del Arca (Uclés, Cuenca)', *Caesaraugusta*, 78: 251–78.
Lakarra, J. 2010. 'Haches, diptongos y otros detalles de alguna importancia: notas sobre numerales (proto) vascos y comparación vasco-ibérica (con un apéndice sobre hiri y otro sobre bat-bi)', *Veleia*, 27: 191–238.
Manniez, Y. 2010. 'Contribution à l'étude des dés en os d'époque romaine de la cité de Nîmes (F)', *Instrumentum*, 32: 18–22.
Michelena, L. 1987. 'Baskisch = Hispanisch oder = Gallisch?', *Veleia*, 2.3, special issue: *Studia palaeohispánica: actas del IV Coloquio sobre lenguas y culturas paleohispánicas*: 95–305.
Moncunill Martí, N. 2007. 'Lèxic d'inscripcions ibèriques (1991–2006)' (unpublished doctoral dissertation, Universitat de Barcelona).
Moret, P. 2008. 'Recherches historiques et archéologiques sur l'Ibérie Antique. Volume 2. Architecture, urbanisme et organisation du territoire dans l'Ibérie de l'âge du Fer et de l'époque républicaine (VIIe–Ier s. av. J.-C.)' (unpublished 'mémoire d'habilitation', University of Toulouse-Le Mirail).
Moret, P., C. Ruiz Darasse, and G. Verrier. 2015. 'Ibère, grec et latin à Toulouse à la fin du IIe s. av. J.-C.: nouvelles inscriptions sur céramique du site de la ZAC Niel', *Gallia*, 72.2: 403–16.
Orduña Aznar, E. 2005. 'Sobre algunos posibles numerales en textos ibéricos', *Palaeohispánica*, 5: 491–506.
———. 2011. 'Los numerales ibéricos y el protovasco', *Veleia*, 28: 125–39.
———. 2019. 'The Vasco-Iberian Theory', in A. G. Sinner and J. Velaza (eds), *Palaeohispanic Languages and Epigraphies* (Oxford: Oxford University Press), pp. 219–40.
Oroz Arizcuren, F. 1979a. 'Dos inscripciones metrológicas ibéricas con posibles elementos "vascos"', *Fontes linguae Vasconum*, 33: 423–44.

———. 1979b. 'El sistema metrológico de la inscripción ibérica del cuenco de La Granjuela', in A. Tovar (ed.), *Actas del II Coloquio sobre lenguas y culturas prerromanas de la Península Ibérica* (Salamanca: Acta Salmanticensia), pp. 283–370.

Pappa, E. 2017. 'The Metrological Systems of the Final Bronze Age Balance Weights and the Pre-Roman Coinage of Atlantic Iberia: A Shared Syrian Standard?', *Journal of Ancient History and Archaeology*, 6.1: 60–80.

Peñalver, X., and S. San Jose. 2011. *La Edad del Hierro en Gipuzkoa* (Donostia-San Sebastián: Arkeologia).

Poigt, T. 2015. 'Le lot de dix poids d'El Cigarralejo (Murcie). Approche de la métrologie ibérique à partir d'un exemple archéologique', *Pallas*, 97: 133–58.

———. 2022. *De poids et de mesure: les instruments de pesée en Europe occidentale durant les âges des Métaux (XIVe–IIIe s. a.C.); conception, usages et utilisateurs* (Pessac: Ausonius éditions).

Poigt, T., and L. Delanaye. 2020. 'Metrological Modifications on Weights from the Iberian Peninsula (6th c. – 1st c. BC): A Computer Based-Approach', in C. Doyen and L. Willocx (eds), *Pondera antiqua et mediaevalia*, 1: *Actes du Workshop Pondera 2, Louvain-la-Neuve, 2017* (Louvain-la-Neuve: Presses universitaires de Louvain), pp. 139–62.

Rodríguez Ramos, J. 2004. *Análisis de epigrafía íbera* (Vitoria-Gasteiz: Servicio Editorial, Universidad del País Vasco).

———. 2005. 'Introducció a l'estudi de les inscripcions ibèriques', *Revista de la Fundació Privada Catalana per a l'Arqueologia Ibèrica*, 1: 13–146.

Ruiz Zapatero, G. 2012. *Los últimos carpetanos: el oppidum de El Llano de la Horca en Santorcaz, Madrid* (Madrid: Museo arqueológico regional).

Sanmartí J., E. Gili, A. Rigo, and J. De la Pinta (eds). 1992. *Els promers pobladors de Santa Coloma de Gramenet: dels origens al mon roma* (Santa Coloma de Gramanet: Museu Torre Balldovina).

Schulten, A. 1931. *Numancia: Die Ergebnisse der Ausgrabungen, 1905–1912* (Munich: Bruckmann).

Torija López, A. 2003. 'Algunas consideraciones para el estudio de la epigrafía ibérica sobre vajilla de plata: el cuenco del Alcornocal', *Palaeohispánica*, 3: 167–78.

Tovar, A. 1955. 'Inscripción ibérica en una gamella del tesoro de La Granjuela', *Revista de archivos, bibliotecas y museos*, 61.2: 580–83.

Valeri, V. 1999. 'Los numerales en las lenguas del área mediterránea', in F. Villar Liébana and F. Beltrán Llóris (eds), *Pueblos, lenguas y escrituras en la Hispania prerromana: actas del VII coloquio sobre lenguas y culturas paleohispánicas* (Salamanca: Ediciones Universidad de Salamanca), pp. 651–62.

Vidal, J. C. 2017. 'La palabra ibérica abaŕ "diez; decena" y el sistema decimal a-o-ki', in J. Aparicio Pérez and L. Silgo Gauche (eds), *XXX seminario de Lenguas y epigrafía antiguas: ponencias y estudios varios* (Valencia: Real Académia de Cultura Valenciana), pp. 107–65.

Vilaça, R. 2003. 'Acerca da existência de ponderais em contextos do Bronze Finale/Ferro Inicial no território português', *O arqueólogo português*, 4th ser., 21: 245–88.

———. 2011. 'Ponderais do Bronze Final-Ferro Inicial do Ocidente Peninsular: novos dados et questões em aberto', in M. P. García-Bellido, L. Callegarin, and A. Jiménez Diéz (eds), *Barter, Money and Coinage in the Ancient Mediterranean (10th–1st Centuries BC)* (Madrid: Consejo Superior de Investigaciones Científicas), pp. 139–68.

Villaronga, L. 1973. 'Marcas de valor en monedas ibéricas', in *XII Congreso Nacional de Arqueología (Zaragoza 1973* (Zaragoza: Universidad de Zaragoza, Seminario de arqueología), pp. 531–36.

Corpus

BDH, Hesperia, Banco de datos de lenguas paleohispánicas <http://hesperia.ucm.es> [accessed 8 April 2022].

LORENZ RAHMSTORF

8. Highly Composite Numbers and the Early Use of Weights

ABSTRACT The concept of measuring mass by weighing emerged in the late fourth and early third millennia BC in Egypt and Mesopotamia. Only a few hundred years later weights are also known in the Aegean and Anatolia in the west and the Indus region in the east. Divisibility is an important characteristic in practical weight usage. Highly composite numbers (like four, six, twelve, twenty-four, thirty-six, forty-eight, sixty, 120, etc.) are positive integers with more divisors than any other positive integers have. Weight values of weights from the sites of Khafaja, Tell Asmar, and Ishchali in central Mesopotamia dating to the third and early second millennia BC are investigated in this contribution in regard to their divisibility. The Sumerians used the excellent advantages of the highly composite number sixty for metrology in order to achieve maximal divisibility when weighing out precise amounts of products like metals.

KEYWORDS Mesopotamia, Late Copper Age, Bronze Age, numbers, sexagesimal system, weights

Introduction

A sexagesimal system means a system of numeration or numerical notation which rises in powers of sixty. When Marvin A. Powell asked in 1973 'why did the Sumerians count in multiples of sixty' he was not able to give an answer (Powell 1973, 81). Still today, the answer is elusive. A likely guess is that this resulted from astronomic observations: 'the adoption of this system is linked up, among other things, to the need to represent astronomical data like the duration of the year (about 360 days)' (Torres-Heredia Julca 2007). The understanding of a month of thirty days and a year of 360 days is documented already in the Uruk III period in the numerical system U4 (Englund 1998, 121, fig. 41). The Uruk IV and III periods are traditionally dated in the last three centuries of the fourth millennium BC (Englund 1998, 23, fig. 2) but recent ^{14}C dates imply a slightly earlier date (van Ess 2015).

While astronomical observations were surely significant for humans since the beginnings of mankind's history and even increased with the beginning of farming and the importance of observing the calendar, this does not really explain the prominence of sixty. The properties of the number are special. Theon of Alexandria (*c.* AD 335–405) already described the

Lorenz Rahmstorf University of Göttingen. Lorenz Rahmstorf is Director and Professor at the Seminar für Ur- und Frühgeschichte at the Georg-August-University of Göttingen, Germany. He has academic degrees from the Universities of Bristol (MA), Heidelberg (PhD), and Mainz (Habilitation). He has led the ERC Consolidator Project WEIGHTANDVALUE (2015–2022) as the Principal Investigator and has specific research interests in early trade.

Number and Measure in Prehistory: Gift, Exchange, and the Transfer of Goods, ed. by Philippe Della Casa and Aleksander Dzbyński, WSA, 6 (Turnhout, 2024), pp. 107–113.
BREPOLS PUBLISHERS
DOI 10.1484/M.WSA-EB.5.137133

specific qualities of the number sixty — '60 is among all the numbers the most convenient, because, being the smallest among all those which have the most divisors, it is the easiest to handle' (Thureau-Dangin 1939, 95–96 with reference). The aim of this short contribution is to demonstrate that the number sixty works perfectly for weight metrology. This fact may be the reason for the long-lasting impact of the sexagesimal system in ancient Mesopotamian history.

Early Mesopotamian Accounting

The formal pattern of numeration in Sumerian exemplifies the significance of ten and sixty (Powell 1973, 80). The number five has the word meaning of 'a full one' indicating the use of the hand for counting. The etymological meaning of ten is unfortunately unknown. There are signs for ten, twenty, thirty, forty, fifty, and sixty. Sixty is written with just one wedge. Multiples of sixty (3600, 216,000, 12,960,000) are also known as numbers. What was being counted determined the numerical system in the accounting systems reported by the Late Uruk cuneiform tablets (late fourth millennium BC). Sexagesimal accounting systems (Sexagesimal System S, Sexagesimal System S') were used for humans, (dead) animals, animal products, tools, stones, and pots with signs for one, ten, sixty, six hundred, 3600, 36,000 — factors being ten, six, ten, six, ten, hence a combination of a decimal and sexagesimal system (Englund 1998, 118–19). The bisexagesimal systems (Bisexagesimal System B; Bisexagesimal System B*) also used the factor two. In addition, there were nine further numerical systems in the arrangement of Robert Englund. The GAN_2 was used to note area measures and counted field sizes by the factors ten?, six, three, ten, and six. The U_4 System noted one day, ten days, one month, and twelve months with the factors ten, three, and twelve. The ŠE System Š, ŠE System Š', ŠE System Š'', and ŠE System Š* were capacity measures for various kind of grain. The DUG_b System and the DUG_c System were used as capacity measures to count milk, dairy fats, and other products. Finally, the system EN has been understood by Adam Falkenstein, Jöran Friberg, Robert Englund, and others as a system to note weight measures, but this is highly speculative as recently demonstrated again by Vitali Bartash (2019, 12–13). There are no identifiable weight measures in the cuneiform tablets from Uruk IV and III in the second half of the fourth millennium BC.

The earliest attestations in writing of a weight unit — the mina — come from the site of Fara/Šuruppak and Ur during Early Dynastic (ED) I–II (Bartash 2019, 39–43). Nearly all of the limited number of ED I–II texts derive from Ur, which explains the rare attestation of the mina before ED III. In a cuneiform tablet (WF 128) from Fara/Šuruppak, which cannot be dated precisely but is either later ED I or ED II (*c.* twenty-ninth–twenty-seventh centuries BC), the mina (ma-na) is mentioned in a list recording possibly 2 mina of copper but also certain numbers of (ram)goats, and 3 *sila* (capacity unit) of lard (Fig. 8.1; Krebernik and others 2014, 345–46, fig. 14). In a text (UET 2 127) of the same date from Ur, 10, 20, 53, and 60 minas of copper are recorded (Bartash 2019, 40–41). The shekel appears in the cuneiform tablets only by ED III, but this is most likely simply due to the very limited number of texts from ED I and ED II. By the Akkad period (*c.* 2350–2100 BC) a fully sexagesimal weighing system is in use: 60 shekel = 1 mina and 60 mina = 1 talent (Powell 1987–1990, 510).

Finds of Early Weights for the Third Millennium BC

The earliest weights worldwide are datable to *c.* 3100/3000 BC. In the Aegean we have the peculiar shape of spools made in stone, spondylus, and rarely metal (lead, bronze, silver) as early as the twenty-ninth century BC at sites like the small early town of Poliochni (Fig. 8.2: 1; Rahmstorf 2016), or at Çukuriçi Höyük (Horejs 2016) near Ephesos on the east Aegean coast. The site is especially important as it yielded both a spool-shaped weight of Aegean/west Anatolian type (Fig. 8.2: 3) and four dark stone weights of rather Mesopotamian type (Fig. 8.2: 2). In Egypt, there is one weight (Mollat 2007) with the name of the first pharaoh of the First Dynasty, Narmer (Fig. 8.2: 4), but it has no archaeological context like the great majority of all known ancient Egyptian weights. An inscribed weight from the mid-Second Dynasty from Tell el Fara'in (von der Way 1997, 153, 162–63, 204, pl. 70, 9 XXII, 9), ancient Buto, is the earliest stratigraphically dated weight from Egypt (Fig. 8.2: 5).

In Mesopotamia, few weights have been documented as archaeological finds from the ED period (*c.* 3000/2900–2350 BC) so far. The archaeological existence of weights from the Late Chalcolithic Uruk period is a matter of debate. The only site that yielded alleged balance weights of this period is Tepe Gawra. The balance weights come from level VIII, dated to the Late Uruk period. Unfortunately, this level is stratigraphically compromised, and hence the chronology of the finds is dubious (Fig. 8.2: 6; Rahmstorf 2014, 429; 2022, 322, 672, 708). W. Hafford (2019) suggested that some objects from even earlier levels could be weights. However, in our understanding the evidence is not conclusive. One or two potential balance weights from the old excavations at Uruk itself from the (later overbuilt) Anu district are possibly datable to the Jemdet Nasr period (Rahmstorf

Vs.	i	1	⸢1$_b$⸣$^{?}$ **urudu ma-⸢na⸣**	1 Mine Kupfer:
		2	NAM-**nu-kuš$_2$**	PN.
		3	2$_b$ **udu-nita**	2 Widder,
		4	⸢2$_b$⸣ **maš$_2$**	1 Ziegenbock:
		5	ŠUL.ŠUL	PN.
		6	[n$_b$] **udu-nita**	1 Widder:
		7	**abzu**-[x]	PN.
		8	[x] **udu**-[**nita**(?)]	[n] Wid(der?):
		9	[...]	[PN].
	ii	1	1$_b$ **ma-na**	1 Mine (Kupfer):
		2	**munus**-DA-**nu-ku**[š$_2$] ⸢**ašgab**⸣	PN, Lederarbeiter.
		3	3$_b$ **i$_3$-nun** ⸢**sila$_3$**⸣	3 Liter Schmalz:
		4	**me**[**s**]-**u$_4$-ba**	PN.
		5	1$_b$ **ma-n**[**a urudu**(?)]	1 Mine (Kupfer):
		6	**a-nu-zu nağa**[**r**]	PN, Zimmermann.
		7	2$_b$ **uz$_3$**	2 Ziegen
		8	[x] ⸢x⸣ ⸢x⸣	
Rs.	iii		(abgebrochen)	
		1′	⸢n$^{?}$⸣ NIG$_2$.NUMUN.⸢x(.x$^{?}$)⸣	...

2

Figure 8.1. Cuneiform tablet (WF 128) from Fara/Šuruppak (1) and its transliteration and translation (2) (after Krebernik and others 2014).

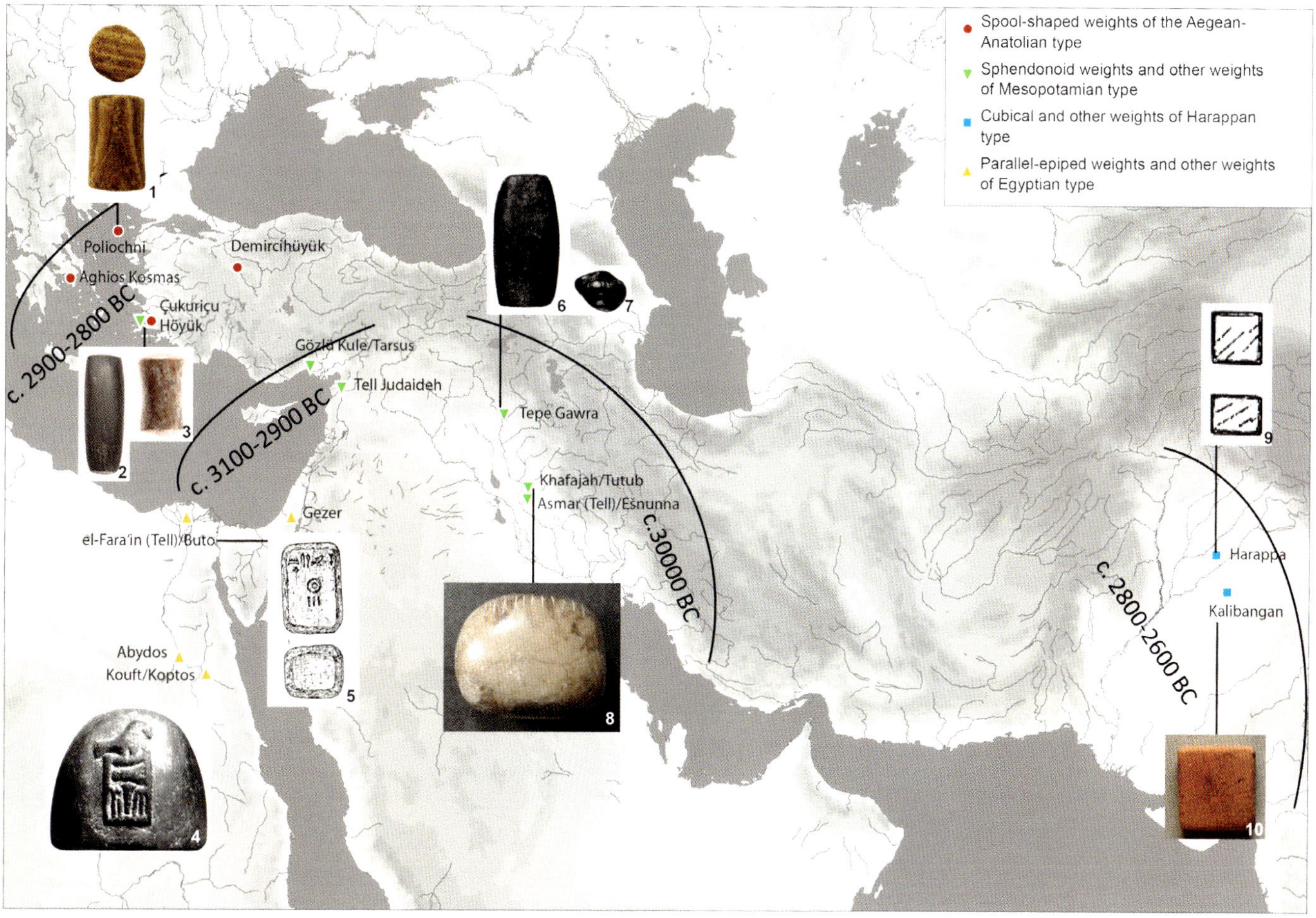

Figure 8.2. Sites with balance weights between c. 3100/3000 and 2600 BC. The figures of the weights are taken from the publications quoted in the text or by author (graphic: Lorenz Rahmstorf).

2014, 429–30; 2022, 344–45, fig. 111.15–16), even if they are sphendonoid hematite weights of a shape typical rather of the late third and early second millennia BC. It is worth reflecting here on the fact that these dubious attestations only come from old excavations, while modern excavations have not yet brought to light any unequivocal weights of the Late Uruk period. Hence, if one considers only the reliable evidence, it seems more appropriate to assume that weighing technology appeared only around 3000 BC with the Early Dynastic period. Definite weights with parallel incised lines (Fig. 8.2: 8) indicating the unit are known from ED I Khafaja, ancient Tutub, in the Diyala region of central Mesopotamia dating to *c.* 3000/2900–2750 BC and again from Tepe Gawra from the Niniveh 5-levels (level VII) of the early third millennium BC. A small duck-shaped object from this level already has the zoomorphic shape which is so typical also for much later Mesopotamian weights (Fig. 8.2: 7).

In the greater Indus region, the first weights of the typical cubic shape can be dated to the late Early Harappan period or Kot Diji phase (*c.* 2800–2600 BC). Specimens of this early phase are only known from a few sites like Harappa (Fig. 8.2: 9), Kalibangan (Fig. 8.2: 10), and Kunal (Rahmstorf 2020, 89; 2022, 451–52, 673, 709). However, only the two exemplars from Harappa can be well dated (Kenoyer 2010, 115). During the Mature Harappan phase (*c.* 2600–1900 BC) more than one thousand weights are known by archaeological finds in Pakistan and western India.

In these various regions, different weight systems were applied which are not yet fully understood for the earliest phases, at least for the Aegean and Egypt. In the Indus, a mixed binary/decimal system (1–2–4; 10–20–40; 100–200–400) was in use (Hemmy 1938; Rahmstorf 2020; 2022, 486–89). In Mesopotamia, the weight system followed a fully sexagesimal system. We will discuss in the following how this was designed and which opportunities this opened up for systematic and precise accounting for the Mesopotamians.

Mesopotamian Weights from the Third and Early Second Millennia

The early history of weight use in Mesopotamia for the ED period (*c.* 3000/2900–2350 BC) is still badly known due to the very few actually known weights from this period. It was one of the principal aims of the ERC-project WEIGHTANDVALUE to improve this situation. Thanks to Helen McDonald and other colleagues at the Oriental Institute at Chicago, Nicola Ialongo and I were allowed to study the still unpublished weights in the Diyala region in eastern Mesopotamia, comprising data from the sites of Khafaja/ancient Tutub, Tell Asmar/ancient Ešnunna, and Ishchali/ancient Nerebtum. The materials cover the whole third millennium and the earlier second millennium BC, and there are interesting chronological differences and few overlaps in between the sites (Delougaz and others 1967). The excavations at Khafaja revealed forty-eight weights mainly of the whole ED period from *c.* 3000/2900 BC to 2350 BC; at Tell Asmar, seventy-eight weights mostly date from late ED to the end of the Akkadian period; and finally, the 183 weights from Ishchali are covering mainly the nineteenth century BC. Of the 309 weights from the three sites, three examples of the most interesting ED/Early Akkadian weights are briefly presented here, all coming from Khafaja, which is well known for its Oval Temple (Fig. 8.3: 1; Delougaz 1940). The shape of the ED weights is often unremarkable, generally pebble-like, but some reveal their nature by showing parallel incisions. One weight, one from ED I (of the earliest known from Mesopotamia) bears eleven incisions (Fig. 8.3: 2 = Fig. 8.2: 8). Another object is cylindrical (Fig. 8.3: 3), and comes from the ED IIIa layers of a building north of the Oval Temple (Fig. 8.3: 4). The third weight is a rounded parallelepiped (Fig. 8.3: 5), discovered in an Early Akkadian building layer east of the Oval Temple (Fig. 8.3: 6). All the data will be published in a detailed study by Nicola Ialongo and me. In this contribution, I would like to demonstrate their relevance of the problem of the emergence of the fully sexagesimal system in Mesopotamian accounting.

The Number Sixty and its Use in Mesopotamian Weight Metrology

Sixty is a highly composite number, meaning that it has more prime factors (i.e. prime numbers that can be multiplied to give the original numbers) than any smaller number (Table 8.1). Highly composite numbers were first systematically investigated by the Indian mathematician Srinivasa Ramanujan ([1915] 1997) in the early twentieth century. Sixty has the prime factors one, two, three, four, five, six, ten, twelve, fifteen, twenty, thirty, and sixty. It is the smallest number divisible by every number from one to six and no smaller number is divisible by the numbers one to five. With twelve prime factors, the number sixty has even more than the number one hundred, which has only nine. Sixty is even a *superior highly composite number* with its specific high divisibility and its integration in various numeral systems, e.g. binary (base two), senary (base six), decimal (base ten), duodecimal (base twelve).

In the cuneiform texts, fractions of half, one-third, and two-thirds of a mina are mentioned. According to the recent examination of V. Bartash, the fractions one-fourth and one-fifth did not exist in the third

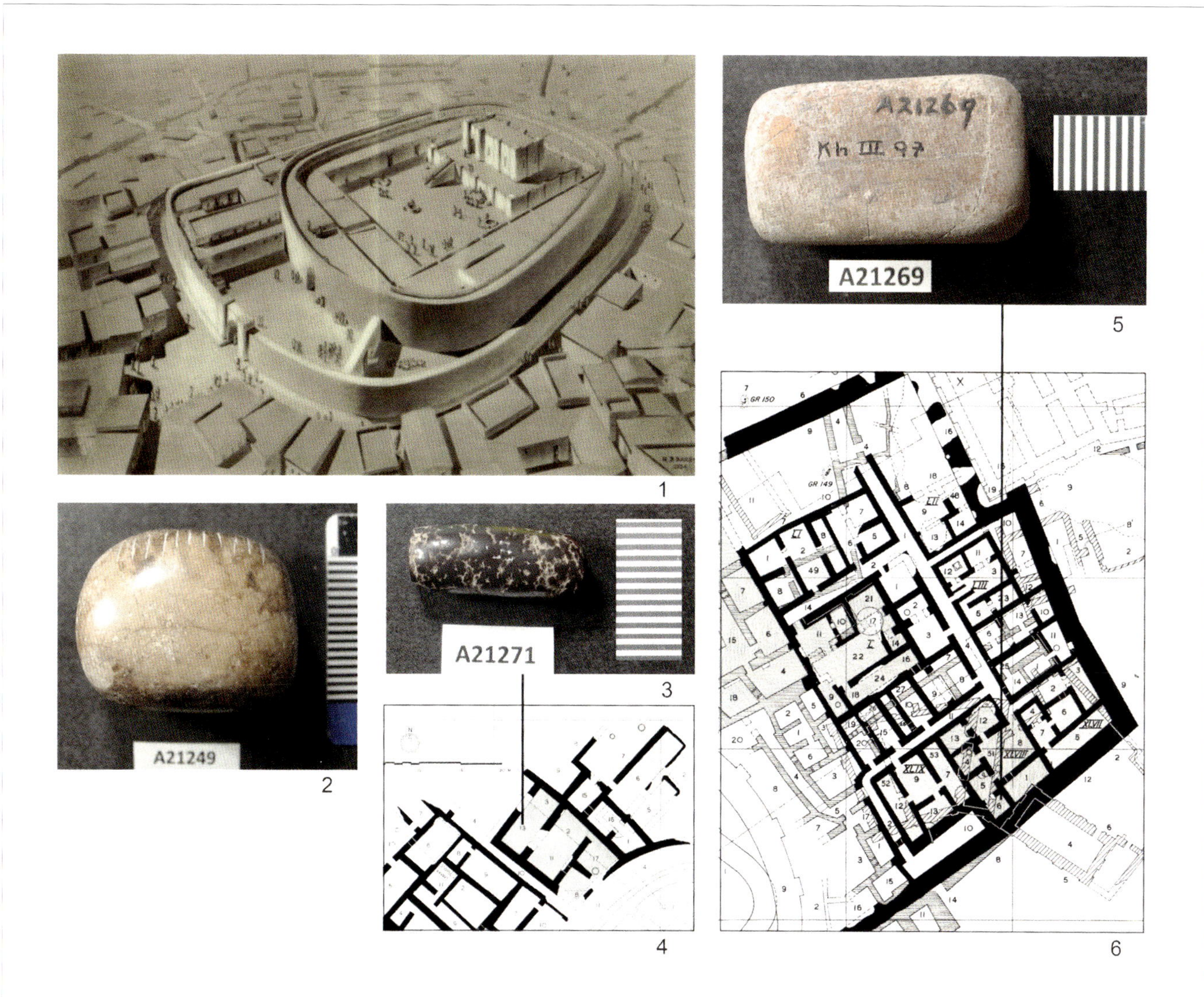

Figure 8.3. Reconstruction of the Temple Oval at Khafaja, Iraq (1), and weights from Early Dynastic (ED) I (2), ED IIIa (3), and Early Akkadian (5) and the architectural context (4, 6) in which they were found (1, 4, and 6 are taken from Delougaz 1940; Delougaz and others 1967. Graphic: Lorenz Rahmstorf).

Table 8.1. Highly composite numbers and the number of divisors.

Number	2	4	6	12	24	36	48	60	120	180	240	360	720	840
Divisors	2	3	4	6	8	9	10	12	16	18	20	24	30	32

Table 8.2. Fractions, divisors, and resulting numbers used in the sexagesimal system. The fractions of 1/4 and 1/5 are not documented in third-millennium cuneiform texts.

Fraction	**1**	**1/2**	**1/3**	1/4	1/5	**1/6**	**1/10**	**1/12**	**1/15**	**1/20**	**1/30**	**1/60**
Divisor	1	2	3	4	5	6	10	12	15	20	30	60
Number	60	30	20	15	12	10	6	5	4	3	2	1

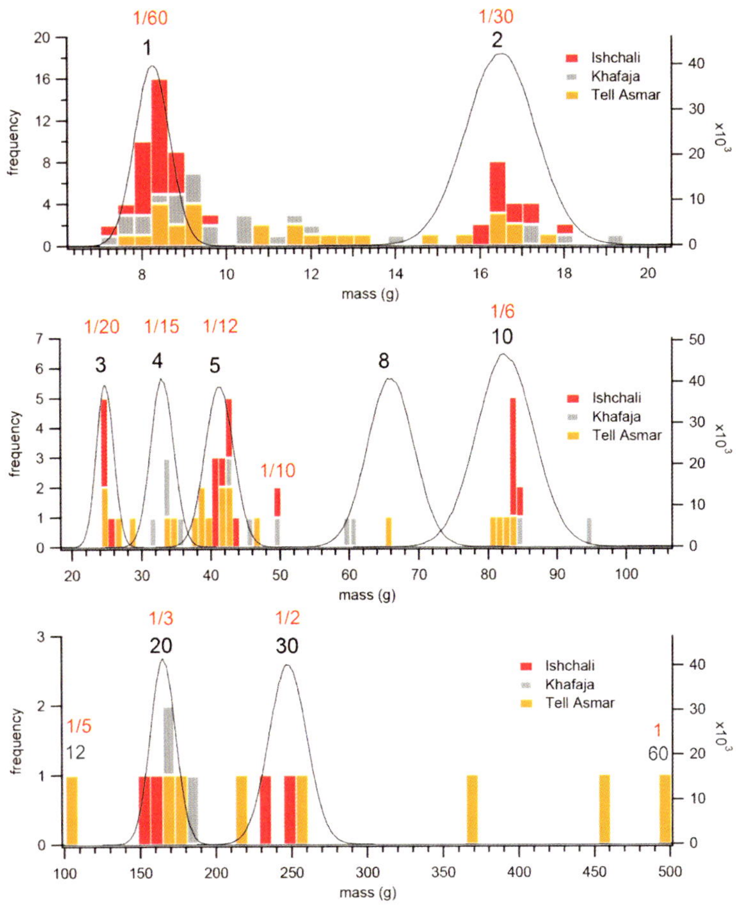

Figure 8.4. Frequency distribution diagram of the weight values of balance weights from Ishchali, Khafaja, and Tell Asmar, Iraq, from the third and early second millennia BC (n = 307) (graphic: Nicola Ialongo).

millennium. Below that weight values were counted in shekels of 10, 5, 4, 3, 2, 1 — such weights are documented in third-millennium BC administrative texts (Bartash 2019, 43, 126). Table 8.2 gives an overview.

The Numerical Configuration of the Weight Values from the Diyala Sites

A frequency distribution diagram of the weight values from Khafaja, Tell Asmar, and Ishchali shows clear peaks at certain values. Most weights from all three sites cluster between *c.* 7.9 g and 8.7 g (Fig. 8.4: top row left). This value is easily identifiable as the Mesopotamian shekel with *c.* 8.3 g as the average value. There are also clear peaks at multiples of this value: two, three, four, five, ten, twenty, and thirty times the unit (Fig. 8.4). The values of the weights from Khafaja and Tell Asmar from the third millennium BC are slightly more variable than the ones from Ishchali in the early second millennium. This is visible, for instance, in the values between ten and fourteen which do not fit to the Mesopotamian shekel. Overall, the distribution of weight values clearly mirrors the sexagesimal Mesopotamian weighing system. All peaks in Figure 8.4 are multiples of the shekel and at the same time fractions of the mina: 1/60 (= 1 shekel), 1/30 (= 2 shekels), 1/20 (= 3 shekels), 1/15 (= 4 shekels), 1/12 (= 5 shekels), 1/10 (= 6 shekels), 1/6 (= 10 shekels), 1/5 (= 12 shekels), 1/3 (= 20 shekels), and 1/2 (= 30 shekels). Of all possible fractions, only 1/4 (15 shekels) is not visible in the sample. 1/5 is present only by one weight. However, this measurement is isolated, and can simply represent an outlier, i.e. a deviant weight generated by chance. As mentioned above, both 1/4 and 1/5 are not recorded in the cuneiform tablets. In the weight sample from the Diyala sites, ten of the possible eleven fractions of sixty were used by specimen of weights. The high divisibility of sixty was therefore clearly used and represented a big advantage of this metrological system over others (e.g. capacity systems) which were much more complicated.

Conclusions

The use of sixty in the Mesopotamian accounting system is most likely very old. It was important in four of the thirteen accounting systems of the Late Uruk period in the later fourth millennium BC. The extreme high divisibility of sixty was, however, not used during this period. It seems that this specific quality was only understood during the third millennium BC when a fully sexagesimal system was developed in weight metrology. The third- and early second-millennium balance weights from the sites Khafaja, Tell Asmar, and Ishchali in the Diyala region of central Mesopotamia demonstrate this very well by physical evidence. The high divisibility allowed for a maximum of potential fractions and calculations. Apparently, it was only through the medium of weight metrology that the sexagesimal system became so successful in Mesopotamia.

Acknowledgements

This work was supported by ERC-2014-CoG 'WEIGHTANDVALUE: Weight Metrology and its Economic and Social Impact on Bronze Age Europe, West and South Asia' (grant no. 648055). My thanks go to Nicola Ialongo, postdoctoral researcher in same project, for calculating and designing Figure 8.4.

Works Cited

Bartash, V. 2019. *Establishing Value: Weight Measures in Early Mesopotamia*, Studies in Ancient Near Eastern Records, 23 (Berlin: De Gruyter).

Delougaz, P. 1940. *The Temple Oval at Khafajah*, Oriental Institute Publications, 53 (Chicago: University of Chicago Press).

Delougaz, P., H. D. Hill, and S. Lloyd. 1967. *Private Houses and Graves in the Diyala Region*, Oriental Institute Publications, 87 (Chicago: University of Chicago Press).

Englund, R. K. 1998. 'Texts from the Late Uruk Period', in J. Bauer, R. K. Englund, and M. Krebernik (eds), *Mesopotamien: Späturuk-Zeit und Frühdynastische Zeit; Annäherungen*, 1, Orbis Biblicus et Orientalis, 160.1 (Freiburg, Göttingen: Vandenhoeck & Ruprecht), pp. 15–233.

Ess, M. van. 2015. 'Absolute Chronology of the Uruk and Jemdet Nasr Periods at Uruk, Southern Mesopotamia: The Interpretation of Additional 14C Samples', *Zeitschrift für Orient-Archäologie*, 8: 10–37 [with a contribution by K.-U. Heußner].

Hafford, W. B. 2019. 'Accounting for Civilization: Early Weights and Measures at Tepe Gawra', in L. Rahmstorf and E. Stratford (eds), *Weights and Marketplaces from the Bronze Age to the Early Modern Period: Proceedings of Two Workshops Funded by the European Research Council (ERC)*, Weight & Value, 1 (Kiel: Wachholtz), pp. 15–34.

Hemmy, A. S. 1938. 'System of Weights', in E. J. H. Mackay, *Further Excavations at Mohenjo-daro: Being an Official Account of the Archaeological Excavations at Mohenjo-daro Carried out by the Government of India between the Years 1927 and 1931* (New Delhi: Government of India Press), pp. 601–12.

Horejs, B. 2016. 'Neue Gewichtssysteme und metallurgischer Aufschwung im frühen 3. Jahrtausend – ein Zufall?', in M. Bartelheim, B. Horejs, and R. Krauß (eds), *Von Baden bis Troia: Ressourcennutzung, Metallurgie und Wissenstransfer; Eine Festschrift für Ernst Pernicka*, Oriental and European Archaeology, 3 (Rahden: Leidorf), pp. 251–72.

Kenoyer, J. M. 2010. 'Measuring the Harappan World: Insights into the Indus Order and Cosmology', in I. Morley and C. Renfrew (eds), *The Archaeology of Measurement: Comprehending Heaven, Earth and Time in Ancient Societies* (Cambridge: Cambridge University Press), pp. 88–105.

Krebernik, M., H. Steible, and F. Yildiz. 2014. 'Prä-Fara-zeitliche Texte aus Fara', in N. V. Koslova, E. Vizirova, and G. Zolyomi (eds), *Studies in Sumerian Language and Literature: Festschrift Joachim Krecher*, Bael und Bibel, 8 (Winona Lake: Eisenbrauns), pp. 327–82.

Mollat, H. 2007. 'Zur Metrologie der Gewichtsstücke Alt-Ägyptens', *Maß und Gewicht*, 84: 2045–77.

Powell, M. A. 1973. 'Sumerian Numeration and Metrology' (unpublished doctoral dissertation, University of Minnesota).

———. 1987–1990. 'Masse und Gewichte', in *Reallexikon der Assyriologie und Vorderasiatischen Archäologie*, VII (Berlin: De Gruyter), pp. 457–530.

Rahmstorf, L. 2014. 'Early Balance Weights in Mesopotamia and Western Syria: Origin and Context', in P. Bieliński, M. Gawlikowski, R. Koliński, D. Ławecka, A. Sołtysiak, and Z. Wygnańska (eds), *Proceedings of the 8th International Congress on the Archaeology of the Ancient Near East, Warsaw, April 30th–May 4th 2012*, III (Wiesbaden: Harrassowitz), pp. 427–41.

———. 2016. 'Emerging Economic Complexity in the Aegean and Western Anatolia during the Earlier Third Millennium BC', in B. Molloy (ed.), *Of Odysseys and Oddities: Scales and Modes of Interaction between Prehistoric Aegean Societies and their Neighbours*, Sheffield Studies in Aegean Archaeology, 10 (Oxford: Oxbow), pp. 225–76.

———. 2020. 'Weight Metrology in the Harappan Civilization', in E. Myrdal (ed.), *South Asian Archaeology and Art 2014* (New Delhi: Dev), pp. 77–96.

———. 2022. *Studien zu Gewichtsmetrologie und Kulturkontakt im 3. Jahrtausend v. Chr.*, Universitätsforschungen zur prähistorischen Archäologie, 379 (Bonn: Habelt).

Ramanujan, S. [1915] 1997. 'Highly Composite Numbers', *The Ramanujan Journal*, 1: 119–53 [original publication from 1915 annotated by J.-L. Nicolas and G. Robin].

Thureau-Dangin, F. 1939. 'Sketch of a History of the Sexagesimal System', *Osiris*, 7: 95–141.

Torres-Heredia Julca, J. V. 2007. 'A Geometrical Link between the Circle and Sexagesimal System', *arXiv*: 0707.0676 [math.HO] <https://doi.org/10.48550/arXiv.0707.0676>.

Way, T. von der. 1997. *Tell el-Fara'în/Buto*, 1: *Ergebnisse zum frühen Kontext. Kampagnen der Jahre 1983–1989*, Archäologische Veröffentlichungen, 83 (Mainz: Von Zabern).

RICHARD SEAFORD †

9. The Functions of Number in Early Greek Texts

ABSTRACT This paper examines numbers in Greek society from the eighth to fifth centuries BC, mainly as reflected in Homer, pre-Socratic cosmology, Aeschylus's Oresteia, and Plato. It focuses on two functions of numbers, both of which are conducive to social integration: on the one hand the creation of traditional memorable forms, which can have a co-ordinating function (notably in ritual), and on the other hand the expression of equivalence, notably in commerce. In the pre-monetary world of Homer the former function is more prominent than the latter. But with the pervasive monetization of Greece through the introduction of coinage (around 600 BC) the use of numbers to express equivalence becomes universal. This new universality of the power of abstract value is in various ways projected by pre-Socratic thought onto the cosmos. In particular, the Pythagorean claim that all things are composed of numbers is a cosmic projection of the all-importance, from a commercial perspective, of numerical price. But the power of number in shaping ritual persists, and from the Oresteia of Aeschylus, who was a Pythagorean, there is a striking example of its synthesis with the cosmic importance for Pythagoreanism of the number three.

KEYWORDS co-ordination, equivalence, Homer, Pythagoreanism, Aeschylus

Practical Functions of Number

I start by distinguishing between two practical functions of number: *co-ordination* of action (singing a verse three times) and *equivalence* (this costs 3 euros). These are of course not the only practical functions of number. Another notable practical function is *differentiation* ('I live at 5 Acacia Road'), which is more common in the modern than in the ancient world (for the variety of human uses of number see Crump 1990).

The functions of number cannot in practice be separated. For instance, *co-ordination* of action may be helped by *differentiation* of time and place, or even by *equivalence* of value (ensuring smooth exchange). *Co-ordination* depends on the number of actions being *memorable* to the participants; but actions may be repeated to create *memorable form* where there is no need for co-ordination — for instance in ritual performed by one person. The effect of an action performed three times may be simultaneously *co-ordination, identifying* itself as a specific *memorable* ritual, the *intensification*

Richard Seaford† University of Exeter. Richard Seaford was Emeritus Professor of Ancient Greek at the University of Exeter. In 2009 he was Honorary President of the British Classical Association. In 2018 a volume of his selected papers was published entitled *Tragedy, Ritual, and Money in Ancient Greece* (Cambridge University Press). His books include commentaries on Euripides' *Cyclops* (Oxford University Press, 1994) and on Euripides' *Bacchae* (Aris and Phillips, 1996), as well as *Reciprocity and Ritual* (Oxford University Press, 1994), *Dionysos* (Routledge, 2006), *Money and the Early Greek Mind* (Cambridge University Press, 2004), *Cosmology and the Polis* (Cambridge University Press, 2012), and *The Origins of Philosophy in Ancient Greece and Ancient India: A Historical Comparison* (Cambridge University Press, 2020).

Number and Measure in Prehistory: Gift, Exchange, and the Transfer of Goods, ed. by Philippe Della Casa and Aleksander Dzbyński, WSA, 6 (Turnhout, 2024), pp. 115–122.
BREPOLS PUBLISHERS
DOI 10.1484/M.WSA-EB.5.137134

inherent in repetition, and the *symbolic power* inherent in the number three.

My limited focus in this paper is on the *merging* of ancient powers of numbers (*co-ordination* and *memorable form*) with their relatively new power of creating *equivalence* that is *universal*. This merging is first manifested in a unique passage of Homer (section 'Homer: Ritual and Equivalence'), and produces in Pythagoreanism a conception of number as ontologically privileged (section 'Pythagoreanism: Number as the Substance of All Things'). It also endows number with the power of social integration, which is evident in Herakleitos (section 'Herakleitos') and Philolaos (section 'Philolaos: Limiting the Unlimited'). This power is exercised especially by the number three (section 'Pythagoreanism: The Integrative Power of Individual Numbers'), which plays a central role in the integration of the *polis* in Aeschylus's *Oresteia* (section 'Three in Aeschylus's *Oresteia*').

Homer: Ritual and Equivalence

In the sacrifice on the seashore in the third book of the *Odyssey* there were nine groups of five hundred men, with nine bulls in front of each (7–8).[1] Here the function of co-ordination is prominent, as often in ritual (for instance when a festival is assigned a set number of days). But when in the eleventh book Odysseus pours three libations to the dead (27–28: 'first with honey-milk, then pleasant wine, the third with water'), the effect of the number three is not *co-ordination* but the *intensification* and the *memorable form* of the traditional action. To take another example at random, in ancient Greek *pharmakos* ritual the victim is beaten seven times on the penis.[2] In Vedic texts a specific number of actions required in a ritual is often assigned one or more equivalences (sometimes in order to explain or justify it): e.g. in the: 'Now there are here five utterances [...] fivefold is the sacrifice, fivefold the animal victim, five are the seasons of the year' (*Śatapatha Brāhmaṇa*, I. 5. 2. 16). But this is not a Greek habit.

When in the sixth book of the *Iliad* Glaukos and Diomedes exchange gifts of hereditary hospitality, the poet famously adds: 'Zeus took away the mind of Glaukos, who exchanged armour with Diomedes, gold for bronze, a hundred oxen's worth for nine oxen's worth' (236). The traditional practice of gift exchange (predominant in Homer), in which numerical equivalence of value does not feature, is seen from the new perspective of commercial exchange (marginal in Homer), in which numerical equivalence of value has become all-important. Glaukos, because he gives something worth a hundred oxen in exchange for something worth only nine oxen, is said (from a commercial perspective) to have lost his mind. In its prefiguring of commercial exchange this passage is unique in Homer: it is the only one that contains the numerical value of both items exchanged (Seaford 2003).[3]

Moreover, the numerical expressions of value in this passage (nine oxen and a hundred oxen) are *taken from sacrificial practice*. Nine and a hundred were both typical numbers of oxen sacrificed. And so the functions of these numbers of oxen in ritual (*co-ordination*, *memorable form*, *intensification*) have — in the gift exchange between Glaukos and Diomedes — been replaced by the new function of *equivalence*, even though the point is to express the lack of equivalence of the items exchanged.

The passage describes a traditional gift exchange (in which numbers have no place) from the new commercial perspective of numerical equivalence of value, but the numerical values are abstracted from the stereotypical numbers of oxen offered to the gods in traditional ritual. Just as traditional gift exchange is here merged with the new dominating reality of commercial exchange, producing a composite event in which Glaukos loses his mind, so the ritual functions of number are merged with the new dominating reality of number as expressing (commercial) *equivalence*.

Pythagoreanism: Number as the Substance of All Things

With the rapid and pervasive monetization of the Greek *polis* as a result of the advent of coinage, from the beginning of the sixth century BC onwards, we would expect a revolution in the conception of number. This is because whereas previously the value of a commodity might be expressed numerically in terms of another commodity (e.g. silver or barley), the *only* use of coinage is as money (i.e. it is not a commodity), and moreover the metallic value of coinage is distinct from (almost always less than) its conventional value. A coin and a jar of oil each embodies a number of *purely abstract units* (2 drachmas). In addition, whereas in ritual

1 Ancient texts are generally available in various editions and translations, notably in the comprehensive Loeb and Budé collections, which provide the Greek text with facing translations in English and French respectively.

2 Tzetzes, *Chil.*, V. 728 ff. on Hipponax fr. 5.

3 Expressed by the (repeated) juxtaposition of antithetical terms (χρύσεα χαλκείων, ἑκατόμβοι> ἐννεαβοίων), a form of expression much commoner in later texts such as Herakleitos and tragedy than it is in Homer.

number is — whether for *co-ordination, intensification,* or *memorable form* — never imagined separately from the sights and sounds in which it is embodied, from the perspective of a trader what matters most about his jar of oil is an abstract *number,* the numerical expression of its value (its identity, quality, size, state, etc. matter *only* insofar as they impact on the number).

We would expect as a result the novel idea that what matters most about an entity is a number that inheres in it but can be imagined entirely separately. And that is precisely what we find, shortly after the monetization of southern Italy, in Pythagoreanism. Monetization takes the equivalence function of number to a new plane, in which number (a) can be imagined *separately* from what it inheres in, indeed it often *must* be so imagined, and (b) often must be imagined as more important or fundamental than what it inheres in, as the *essence* of what it inheres in. According to the best source for early Pythagoreanism, Aristotle (Burkert 1972), number was regarded as the οὐσία (being, substance, essence) of all things (Seaford 2004, 270).[4] This does not mean however that the Pythagoreans had the conceptual means to separate the numbers into a metaphysical realm of their own, as Plato subsequently did (Burkert 1972, 31).

It is astonishing that in the vast scholarly literature on Pythagoreanism monetization is almost never mentioned. The exception is George Thomson, who made the connection without noticing that it was in effect already made in antiquity (1961): Aristoxenos, who in fourth-century BC Tarentum was well acquainted with Pythagoreanism (Minar 1942; Burkert 1972, 107–08), stated that 'Pythagoras seems to have honoured, most of all, the study of numbers, and to have advanced it in diverting it from the use of traders, likening all things to numbers'.[5] There is abundant further evidence for this connection (as more important than the connection with musical harmony), which I have set out elsewhere as part of an argument that pre-Socratic cosmology as a whole projects the new universal power-substance of money onto the cosmos (Seaford 2004).

Walter Burkert in his book on Pythagoreanism gave various cross-cultural examples of *number symbolism,* i.e. of numbers having associations (e.g. three with male, four with female) and as a means of creating order and correspondence in the cosmos. This phenomenon is indeed important for understanding Pythagoreanism. But Burkert (1972, 477) claims that 'the Pythagorean doctrine that "all is number" grows directly out of "archetypal" number symbolism'. The word 'directly' misleads. The Pythagorean idea that (all) things *consist of* numbers is not found in the comparative data, and is quite different from — indeed inconsistent with — the primitive phenomenon, adduced by Burkert, of each number having 'its own individual physiognomy' (Seaford 2004, 271). Certainly unprecedented in Pythagoreanism is that numbers combine perfect abstraction with ontological privilege.

The function of number in monetized exchange is *equivalence,* which may — as noted above (section 'Practical Functions of Number') — enable *co-ordination.* But whereas the number of actions or utterances in a (ritual or non-ritual) performance may help to co-ordinate the pre-existing unanimity of a group, the pre-existing attitude of the parties to a commercial exchange (*qua* commercial exchange) is absolute opposition: each party desires to give as little as possible and receive as much as possible. Number therefore, embodied in money in co-ordinating exchange, may seem to unite opposites. This brings us to an Ionian contemporary of Pythagoras, Herakleitos.[6]

Herakleitos

For Herakleitos 'all things are an exchange for fire and fire for all things, like goods for gold and gold for goods' (B90).[7] I have argued in detail for the importance of money as a factor in Herakleitos's imagining of the cosmos (Seaford 2012, 231–44). The process of exchange of all things for fire is *measured.* 'This cosmos, the same of all things, did none of the gods or humans make; but it was always and is and will be an ever-living fire, kindling in measures and going out in measures' (B30). For instance, '<earth> is dispersed as sea, and is measured so as to form the same *logos* as existed before it became earth' (B31), where *logos* seems to mean something like a numerical proportion. When used of quantity, *logos* means quantity expressed as an abstraction, and could in the fifth century BC (and probably earlier) mean a monetary account (Seaford 2004, 231). All things, says Herakleitos, happen according to the *logos* (B1). Just as the circulation of goods is driven by all-underlying money according to a numerical abstraction (*logos*) embodied in the money, so cosmic circulation is driven by all-underlying fire according to the *logos* embodied

4 *Metaphysics,* 987a19. There are in fact various ways in which Aristotle describes the relationship between numbers and things in Pythagoreanism.

5 DK 58B2 = B23 Wehrli. Traders may of course use numbers without using money. But what was radically new in the time of early Pythagoreanism was both its conception of number and monetization.

6 Pythagoras too was an Ionian, who migrated to southern Italy from Samos, and is mentioned by Herakleitos (B40).

7 All references to pre-Socratic philosophy are to Diels and Kranz 1951.

in the fire (Seaford 2004, 232–33). The *logos*, like the value of money, is communal (B2, B114).

Scholars have been puzzled by Herakleitos's statement that 'one should know that war is communal, and justice is conflict, and all things happen according to conflict and necessity' (B80). Puzzlement dissolves only when we realize that Herakleitean cosmology sublimates monetized exchange, which is universally ruled not only by the necessity inherent in monetized debt, a necessity which is imagined as having impersonal force, but also by the conflict latent in diametrically opposed interests that are brought into harmony by communally agreed but impersonal abstract (numerical) equivalence (the *logos*), i.e. in effect by number: and so conflict is universal, communal, and just, and opposites are held together in unity or 'harmony' (B8, B51). The circulation of money is in Herakleitos *universalized*: the all-pervasiveness of money becomes cosmic, debt becomes universal necessity, numerical formula (*logos*) becomes a universal abstract principle that eventually becomes 'reason', and opposition inherent in monetized exchange becomes the universal unity of opposites.

Philolaos: Limiting the Unlimited

'Although the *logos* is communal', says Herakleitos, 'the many live as if having individual understanding' (B2). Although money works only by communal agreement, it *individualizes*, and from Solon onwards the Greeks are conscious that the circulation and accumulation of — and individual desire for — money is inherently *unlimited*, and that this unlimitedness is dangerous (Seaford 2004, 165–69). For Solon as lawgiver the problem was that 'of wealth there is no limit that appears to men. For those of us who have the most wealth are eager to double it' (fr. 13. 71–73). Accordingly, he projects the necessary limit onto the cosmos: 'it is very difficult to understand the invisible measure of intelligence, which alone holds the limits (πείρατα) of all things' (fr. 16). Similarly, for Herakleitos 'invisible harmony is stronger than visible' (B54).

Given that money and number are fundamental to Pythagorean metaphysics, so too are limit, the unlimited, and harmony. The earliest surviving Pythagorean texts are the six or seven genuine fragments of Philolaos of Croton (or Tarentum, *c.* 470–390 BC). Having number is a precondition for being known (B4). Existing things are either limiting or unlimited or both (B2), and cosmogony was a fitting-together of unlimiteds and limiters (B1, B6), as a result of harmony coming upon them (B2, B6). In his detailed study of Philolaos, Huffman (1993, 73) concludes that he 'seems [...] to conceive of all "fitting together" of limiters and unlimiteds in terms of numerically specific relations'. The nature of these abstract entities is mysterious. Why did Philolaos make them fundamental?

Without giving a full explanation here, I note merely that the harmonization of limiter and unlimited embodies the newly fundamental social relation, monetized exchange (see further Seaford 2004, 275–83). Aristotle notes that currency is a *limit of exchange* (*Politics*, 1257b23), before saying that all those engaged in the art of acquisition increase their currency without limit (1257b34). In what sense then does currency limit exchange? As Reeve (1998, 17) comments, on his translation of the passage, 'the price of something delimits its exchange value'. Payment of 5 drachmas for a vase is an agreement to separate, in a single exchange, a limited sum of money from the unlimited sum of (homogeneous) money in unlimited circulation.[8] In reality the price is not given by the vase itself: the amount of money that *could* be given for the vase is unlimited. And indeed it is in the interest of the seller to increase the limit as much as possible, of the buyer to reduce it as much as possible. The price actually paid depends, *we* would say, on various factors that may include supply and demand, the amount of labour involved in making the vase, and the determination and bargaining skill of the parties to the exchange.

However, in this newly monetized society this theoretical perspective is not available. Instead, two unprecedented but fundamental features of money predominate. One is its necessary *impersonality* (unlike the gift, it must be entirely separable from its owner), the other is its *abstractness*. For Herakleitos the cosmic cycle, imagined in terms of the monetized cycle of exchange, occurs according to a *numerical* formula (*logos*) that is *impersonal* and *abstract*. What determines price has to be imagined as *impersonal* (*objective*), beyond the (disruptive) will of the transactors. For Philolaos too the cosmic process, imagined in terms of the monetized transaction, involves impersonal and abstract entities (unlimiteds, limiters, harmony). In reality everything depends on agreement to a price, which is in the metaphysical imagination achieved by the power of ἁρμονία (harmony) to conjoin unlimited (the interest of the seller) with limiter (the interest of the buyer). Whereas in performance the co-ordinating function of number helps enactment and reinforcement of an existing unanimity (three verses, three circuits of the altar), in commerce it helps to create unanimity out of opposition.

8 This form of delimitation distinguishes the monetary use of number from other of its uses that may — like human culture in general — be said to be delimiting (choosing three men to hunt bison delimits the three men from all the others and from each other).

But this by itself is not sufficient to explain the importance of limiting the unlimited in Pythagorean cosmogony. The most important source of this importance was, as it was for Solon, surely the disruptive power of the unlimitedness of money (of its accumulation and of the desire for it) that is nevertheless delimited in every transaction. Philolaos's Tarentine Pythagorean successor Archytas (fl. *c.* 400–350 BC) obligingly supplies an account of some of the lived experience of which Pythagorean metaphysics is a projection. 'The discovery of λογισμός (calculation) ended civil conflict and increased concord. For when there is calculation there is no unfair advantage, and there is equality, for it is by calculation that we come to agreement in our transactions' (47 DK B3). Here too the power that creates agreement (in exchanges and beyond) is numerical, neutral, and impersonal. But Archytas is also a mathematician and a metaphysician. For understanding the cosmos as a whole, 'it seems that logistic (λογιστικά) is with regard to wisdom far superior to the other skills' (B4). Archytas chose — as his name for the entire science of number — λογιστικά, which emphasizes 'the study of relative quantity, i.e. ratio and proportion, rather than the classification of types of number, with which arithmetic (ἀριθμητική) was concerned'. So Huffman, who explains the choice thus: 'λογιστικά, as the study of the relative quality of numbers, is crucial for Archytas in determining a fair distribution of goods and power in the state' (2005, 240, 243).

The *ontological privilege of number* is given neither by nature nor by pure thought. It is an idea that emerges from two aspects of *practice*: firstly from the fact that for traders what matters about goods is their universal numerical (monetary) value; and secondly from the impressive power of number to create concord in prosperity, a power that may also have motivated Herakleitos's advocacy of the communal *logos*. Pre-monetary powers of number, to co-ordinate and to create equivalence, could achieve forms of mediation and integration. But with monetization the power of number to create equivalence becomes universal, and so becomes a means of integrating the *polis*. Thinking in numbers becomes good for the individual and good for the *polis*.

Pythagoreanism: The Integrative Power of Individual Numbers

In Pythagoreanism the (money-derived) relative detachment of number from what it inheres in, its abstractness, enables focus on (abstract) inter-numerical *relations* — between numbers or within any number, i.e. what we call pure mathematics. When combined with the Pythagorean (also money-derived) ontological privileging of number, this promotes inter-numerical relations as the basis for the identification of numbers with things. Our best source for early Pythagoreanism is Aristotle (Burkert 1972), who says that for the Pythagoreans 'justice is an affection (πάθος) of numbers'.[9] In his comment on this passage Alexander of Aphrodisias (38. 14–15) says that some identified justice with the number four,[10] 'since being the first square it is divided into equal parts and is equal (for it is twice two)': this identification may have been known to Aristotle, because he states that Pythagoreans called justice simple reciprocity[11] and (if it is Aristotle) a square number.[12] Plutarch records that five has been called marriage, as a union of two (the first even number, associated with the female) and three (the first odd number, associated with the male).[13] This is Pythagorean doctrine, alluded to already by Aristotle.[14] Justice and marriage are each envisaged as a union of opposed entities expressed as number, justice on the basis of equality (two plus two), marriage on the basis of difference, with one entity (odd, male) controlling the other (even, female) (on the ideological roots of this doctrine see Seaford 2012, 284–85). The power of number to create concord, originating in action or performance and then reinforced by its universal power over agreement in monetized exchange, is extended to central social institutions (justice and marriage).

Of particular interest, from this perspective, is the number three. It embodies concord by virtue of its internal relations, by the unity of its components, just as we have seen the numbers four and five doing. Plato in his Pythagoreanizing cosmology in the *Timaios* (31bc) states that

> it is not possible for two things alone to combine well without a third, for it is necessary for there to be some bond in the middle to bring them together. And the fairest bond is the one that most unites itself and the things bound together.

Similarly, in his Pythagoreanizing ontology (Huffman 1993, 106) in the *Philebos* Plato states that limit (πέρας) should control the unlimited, and that the mixture of the finite with the infinite produces a 'third thing' that 'puts an end to the conflict of opposites' (25c–26d).

9 *Metaphysics*, 985b29; cf. also 1078b23.

10 He adds that others identified it as nine (for the same kind of reason).

11 *Nicomachean Ethics*, 1132b23.

12 *Magna moralia*, 1182a11–14.

13 *The E at Delphi*, 388a.

14 Fr. 203 (= Alexander of Aphrodisias, *Commentary on Metaphysics*, XLIX. 8); fr. 199; Aristotle, *Metaphysics*, 1078b23.

Three embodies the unity not of specific opposites (as five does the unity of male and female in marriage) but of unity or wholeness as such. Accordingly, it embodies the unity or wholeness of all things, of the world. 'As the Pythagoreans say, the world and all that is in it is defined (ὥρισται) by the number three, since beginning and middle and end have the number of the whole, and that is what gives the triad', says Aristotle (*De caelo*, 268a10), and continues thus: 'and so, having taken this number from nature as it were one of her laws, we make use of it even for the worship of the gods'. An example of this religious use of three is provided by a Pythagoreanizing interpretation of ritual by Philochorus (328 *FGrH* F87):[15] 'The third krater was for Zeus Τέλειος Σωτήρ (Completing Saviour) because three is the first complete (τέλειος) number, having beginning and end and middle'. Here what we regarded as a fundamental and traditional function of number, of creating *memorable form* in ritual, is attributed to a conception of number that is in fact — we have argued — much more recent.

Three in Aeschylus's *Oresteia*

According to an ancient tradition Aeschylus was a Pythagorean (Seaford 2012, 281–83).[16] On this basis I demonstrated in 2012 that his *Oresteia* contains Pythagorean ideas (Seaford 2012, 281–315). Aristotle records the Pythagorean belief that the fundamental constituents of the world are ten pairs of opposites, of which I showed that six (Limit–Unlimited, Male–Female, One–Many, Still–Moving, Light–Darkness, Good–Bad) are fundamental to the transition dramatized in the *Eumenides*.

Along with these (abstract or general) oppositions in the *Oresteia* there are specific ones, for instance between Orestes and the Erinyes (Furies), and between the violence and the counter-violence. The Fury is described as 'bringing one disaster onto the other disaster' (ἄτην ἑτέραν ἐπάγουσαν ἐπ' ἄτηι) (*Choephoroi*, 402–04). Each act of violence opposes the previous one, but also seems identical to it. 'This insult has come in exchange for insult, | they are hard to tell apart (κρῖναι); he plunders the plunderer'.[17] This identity of opposites is expressed in powerful chiasmus at *Choephoroi* 312–13 ('in return for bloody blow, a bloody blow let one pay'). This opposition-cum-identity may be described as a unity of opposites. In fact the first two plays of the trilogy are full of various instances of the (Herakleitean) unity of opposites (Seaford 2003; 2012, 225–80).

I must here emphasize that the unlimited cycle of violence in the *Oresteia* is associated not only with the unity of opposites but also with the cycle of exchange that produces unlimited accumulation of wealth (I show this in detail at Seaford 2012, 196–205). And so the limiting of the unlimited in *Eumenides* is not just an end to violence but also by implication an end to the kind of unlimited (incipiently monetized) individual enrichment deplored by Solon. The transformed Erinyes sing to the people of Athens (996) 'May you rejoice in your αἰσιμίαι (fated just shares) of wealth'. In this respect limiting the unlimited in the *Oresteia* is of the same kind that — I have argued — was an important factor in the metaphysical privileging of limit over the unlimited in Pythagoreanism.

As for the unity of opposites in the *Oresteia*, I will here confine myself to two instances, each chosen as an aspect of the central opposition between violence and counter-violence.

The *victory* of Agamemnon over Troy is also a *defeat* (for the Trojans). This is a unity of opposites as exemplified by Herakleitos B60 ('the way up and down is the same'). Also, the victory of Agamemnon *becomes* — by implication because of its excesses — his defeat (by Klytaimestra). This is a unity of opposites as exemplified by Herakleitos B67 ('god is day night, winter summer, …').

In *Agamemnon* Klytaimestra is characterized as male (11), Agamemnon as female (918): in each case there is unity of opposites as exemplified by Herakleitos B60. Also, in *Agamemnon* a male is killed by a female, in *Choephoroi* a female by a male: in the sequence of violence and counter-violence the gender relationship is transformed into its opposite (as Herakleitos B67).

What has all this to do with the number three? In the third play of the trilogy Orestes by being acquitted wins a 'victory' (741 νικᾶι) that is stated by Athena to be *not* a defeat for the Erinyes (Furies) (795 οὐ γὰρ νενίκησθε). It is rather the *polis*, and the principle of permanent positive reciprocity, that have won the 'victory'.[18] Similarly, Solon denied 'victory to each side in the civil conflict' (fr. 5. 6), and stood 'like a marker in the middle ground' (fr. 37. 9–190). The key to stable peace is to integrate the two opposed entities by a third — in the *Eumenides* a *polis* institution, the law court created by and presided over by Athena. The Herakleitean world of the first two plays of the trilogy is in various ways replaced by a Pythagorean world in the third (Seaford 2012). 'To every middle (μέσον) god gave power' (*Eumenides* 529).

15 Jacoby 1923–1958.

16 Cicero, *Tusculan Disputations*, III. 23.

17 *Agamemnon*, 1560–61; see also e.g. *Choephoroi*, 461.

18 903 ἀστύνικον [...] πόλιν, 1009 πολέως ἐπὶ νίκηι, 974–75 νίκαι δ' ἀγαθῶν ἔρις.

Now for our other instance of the Herakleitean unity of opposites, of male and female. Athena breaks up this as well. In the trial, in which Apollo and Orestes are on one side and the (female) Erinyes on the other, the acquittal of Orestes rests on the privileging of patriliny over matriliny, and of marriage over physical kinship. Athena arrives in Athens from the battlefield (397–404; cf. 292–96), and may well be wearing armour (as in the *Iliad* and often in vase-painting). And she gives the casting vote for the acquittal of Orestes on the grounds that she has no mother and so approves the male in all things (except for marriage) (735–41). Like Klytaimestra and Agamemnon, Athena combines female with male, and yet this unity of opposites is explicitly resolved, as far as the fundamental issue is concerned, by a female in favour of the male. As a result of her authority as a virgin female deity setting up the law court, and of the honour finally accorded to the Erinyes, the opposition between male (and marriage) and female (and physical kinship), as between victory and defeat, is replaced by consensual *integration*.

And so Athena limits the unlimited (cycle of violence) by breaking up, and rearranging in a Pythagorean way, the Herakleitean unity of violence and counter-violence, victory and defeat, male and female.

To the six pairs of Pythagorean opposites that in 2012 I described in the *Oresteia* I now add a seventh, between odd and even numbers.[19] For another Pythagorean aspect of Athena's achievement involves precisely this opposition. First we must note the importance and appeal of the opposition. To his Pythagorean list of ten pairs of opposites that are the fundamental constituents of the world Aristotle adds (990a8) that 'the only things assumed are limit and unlimit and odd and even' (the first two pairs on his list). Limit is aligned or even identified (Burkert 1972, 32; Guthrie 1962, 242) with odd, in part at least because an odd number contains a middle entity that can integrate (impose a stable relationship between) the two other (equal) numbers of which it is composed, and cannot disintegrate into two even numbers. Moreover, an odd number in union with (added to) an even one remains odd. For instance, limit and odd are aligned with male, even and unlimited with female, and three (male) added to two (female) makes five (marriage): the male middle entity of three (in a later text the penis,[20] Seaford 2012, 284–85) becomes in marriage the integrating middle entity of five. When Athena announces that, because she favours the male, she will vote for Orestes, she adds that Orestes will be acquitted 'even if the votes [of the jury] come out equal' (741), i.e. that her vote goes for acquittal (735). This was the aetiological myth of the Athenian practice of acquitting when the votes are equal (Seaford 1995). A tied vote does not in itself end conflict. The *Oresteia* provides a perfect example — at the very moment of limiting the unlimited — of the potential unlimitedness of even numbers and of the integrating, limiting, stabilizing power of odd numbers.

The triadic action of the *Oresteia* corresponds to the shape of the trilogy, with the first play dramatizing the violence, the second play the counter-violence, and the third play the third-party mediation. The Pythagorean conception of the number three, in which two opposed parties are integrated by a third, is also the number of the whole, of the whole trilogy, but also of the whole universe.

I begin with the triadic *spatial* aspect of the universe. According to Aeschylus (fr. 55) a first libation was poured for Zeus and Hera, the second for the heroes, and the third for Zeus Σωτήρ (Saviour), who in the Pythagoreanizing view of the third krater in Philochorus is associated — we saw — with the completing power of the number three. This threefold libation was common practice, and often evoked in the *Oresteia* (Seaford 2012, 301–02; Burian 1986). Given that the first libation is for Olympians and the second for (chthonic) heroes, the completing power of Zeus Σωτήρ (Olympian *and* chthonic) is to unite the upper and lower worlds. The Zeus for whom Aeschylus's Klytaimestra gives her third blow to Agamemnon is Σωτήρ 'of the dead'.[21] Aeschylus's Danaids pray to 'the gods above and the honoured chthonic ones occupying the tombs and Zeus Σωτήρ the third' (*Suppliants*, 24–26).

As for the temporal aspect of the triadic universe, the parodos of the *Agamemnon* narrates how Zeus is third ruler of the universe after Ouranos and Kronos. Zeus's victory is described as follows: 'And he who next came into being met his τριακτήρ and is gone. Anyone who with good will shouts "Zeus" in victory-song will obtain understanding completely' (171–75). The word τριακτήρ, 'victor by virtue of three throws' (in wrestling), occurs nowhere else.[22] It is connected with τριάζειν or τριάσσειν, which means to win by virtue of three throws, and τριαγμός, which occurs only as the title of a book — by the Pythagoreanizing contemporary of Aeschylus, Ion of Chios — that began with the statement that all things are three (Seaford 2012, 286, 300). The violence of first Kronos replacing Ouranos, and then Zeus replacing Kronos, ends — in a third

19 That leaves only three of the ten Pythagorean opposites unexpressed (so far as I can see) in *Oresteia*: Right–Left, Straight–Bent, Square–Oblong.

20 Plutarch, *On the E at Delphi*, 388ab.

21 *Agamemnon*, 1387; cf. Aeschylus, *Suppliants*, 154–61.

22 There was a wrestling contest between Kronos and Zeus at Olympia (Pausanias v. 7. 10; viii. 2. 2).

stage — with the permanent rule of Zeus, which prefigures the two acts of violence in *Agamemnon* and *Choephoroi* ending — in the third play — with the permanent dispensation presided over by Zeus. Orestes, once acquitted, refers to his salvation as 'by the will of Pallas and of Loxias and of the third Σωτήρ who completes all things' (*Eumenides*, 758–60). Zeus Σωτήρ who receives the third krater in Philochorus (section 'Pythagoreanism: The Integrative Power of Individual Numbers') is also Τελεῖος, 'completing'.

We have seen the power of number to integrate the *polis* by facilitating peaceful exchanges. Another mode of numerical integration of the *polis* occurred in Athens when Aeschylus was eighteen years old: the reforms of Kleisthenes that founded Athenian democracy used number to co-ordinate the *polis* geographically in a manner that has been related to Pythagoreanism (Seaford 2012, 68–71). It was about the same time that tragedy was, not long after the pervasive monetization of Athens, established at the city Dionysia. Tragedy originates in ritual, is performed in a ritual context, and dramatizes numerous rituals. In its use of the number three the *Oresteia* derives from Pythagoreanism (a) the metaphysical centrality of number (derived from its function in creating *equivalence*) as well as (b) the power of pure number (through internal relations) to create *concord*. But it also combines them with (c) the more ancient power of number manifest in *ritual*: its memorable form that may have originated in its power to co-ordinate action.

We saw that in the Homeric account of the exchange of armour between Diomedes and Glaukos the traditional ritual function of number and its new function of (commercial) equivalence combine to reveal their dissonance. Subsequently the commercial function, made all-pervasive by monetization, generates a metaphysics of number that in the *Oresteia* combines with the ancient ritual function of number so as to help end the cycle of revenge driven by the unlimited pursuit of wealth.

Works Cited

Burian, P. 1986. 'Zeus ΣΩΤΗΡ ΤΡΙΤΟΣ and Some Triads in Aeschylus' *Oresteia*', *American Journal of Philology*, 107: 332–42.

Burkert, W. 1972. *Lore and Science in Ancient Pythagoreanism* (Cambridge, MA: Harvard University Press).

Crump, T. 1990. *The Anthropology of Numbers* (Cambridge: Cambridge University Press).

Diels, H., and W. Kranz (eds). 1951. *Die Fragmente der Vorsokratiker*, 6th edn (Berlin: Weidmann).

Guthrie, W. K. C. 1962. *A History of Greek Philosophy*, I (Cambridge: Cambridge University Press).

Huffman, C. 1993. *Philolaos of Croton* (Cambridge: Cambridge University Press).

———. 2005. *Archytas of Tarentum* (Cambridge: Cambridge University Press).

Jacoby, F. (ed.). 1923–1958. *Die Fragmente der Griechischer Historiker*, 3 vols in 15 (Berlin: Weidmann and Leiden: Brill).

Minar, E. L. 1942. *Early Pythagorean Politics* (Baltimore: Waverly).

Reeve, C. D. C. 1998. *Aristotle: Politics* (Indianapolis: Hackett).

Seaford, R. 1995. 'Historicising Tragic Ambivalence: The Vote of Athena', in B. Goff (ed.), *History, Tragedy, Theory* (Austin: University of Texas Press), pp. 202–21.

———. 2003. 'Aeschylus and the Unity of Opposites', *Journal of Hellenic Studies*, 123: 141–63.

———. 2004. *Money and the Early Greek Mind* (Cambridge: Cambridge University Press).

———. 2012. *Cosmology and the Polis* (Cambridge: Cambridge University Press).

Thomson, G. 1961. *The First Philosophers* (London: Lawrence and Wishart).

WARSAW STUDIES IN ARCHAEOLOGY

All volumes in this series are evaluated by an Editorial Board, strictly on academic grounds, based on reports prepared by referees who have been commissioned by virtue of their specialism in the appropriate field. The Board ensures that the screening is done independently and without conflicts of interest. The definitive texts supplied by authors are also subject to review by the Board before being approved for publication. Further, the volumes are copyedited to conform to the publisher's stylebook and to the best international academic standards in the field.

Titles in Series

Bartosz Kontny, *The Archaeology of War: Studies on Weapons of Barbarian Europe in the Roman and Migration Period* (2024)

Zbigniew Polak and Michał Starski, *Archaeological Finds from the Main Town in Gdańsk: A Catalogue from Excavations at Długi Targ-Powroźnicza-Ogarna City Quarter* (2022)

Sympozjum Egejskie: Papers in Aegean Archaeology 3, ed. by Stephanie Aulsebrook, Katarzyna Żebrowska, Agata Ulanowska, and Kazimierz Lewartowski (2022)

Agnieszka Tomas, *Living with the Army II: The Results of Remote Sensing and Fieldwalking Surveys in Novae (Lower Moesia)* (2022)

Sympozjum Egejskie: Papers in Aegean Archaeology 4, ed. by Stephanie Aulsebrook, Katarzyna Żebrowska, and Agata Ulanowska (2023)

Titles in Preparation

Archaeology: Just Add Water: 'Światowit' Supplement Series U: Underwater Archaeology, vol. III, ed. by Bartosz Kontny, Małgorzata Mileszczyk, and Magdalena Nowakowska

Counterfeits, Imitations, and Copies of Roman Imperial Denarii: Making and Faking Coins on Both Sides of the Limes, ed. by Arkadiusz Dymowski and Kyrylo Myzgin